The Mummy And Miss Nitocris
A Phantasy
Of The Fourth Dimension

by

George Chetwynd Griffith

Double 9
BOOKS

The Mummy And Miss Nitocris
A Phantasy
Of The Fourth Dimension
by George Chetwynd Griffith

ISBN: 978-93-59320-17-5

Published by

DOUBLE 9 BOOKS

2/13-B, Ansari Road
Daryaganj, New Delhi – 110002
info@double9books.com
www.double9books.com
Tel. 011-40042856

ABOUT THE AUTHOR

George Chetwynd Griffith-Jones (August 20, 1857 – June 4, 1906) was a British author. He was mostly engaged in the science fiction genre, or scientific romance as it was known at the time, writing many future-war novels and playing a crucial influence in molding that burgeoning subgenre. For a brief period of time, he was the most popular and commercially successful science fiction author in his home nation. Griffith grew up with his parents and older brother, receiving homeschooling and moving frequently as a result of his father's minister vocation. Griffith attended school for a little more than a year after his father died when he was 14 years old before leaving England and traveling the world, returning at the age of 19. He subsequently taught for ten years before embarking on a writing career. After an initial setback that left Griffith unable to support himself, C. Arthur Pearson employed him in 1890. Griffith made his literary debut with The Angel of the Revolution (1893), which was serialized in Pearson's Weekly before being published as a book. He got an exclusive contract with Pearson and followed it up with the equally successful sequel Olga Romanoff (1894).

CONTENTS

CHAPTER I
INTRODUCES THE MUMMY

"Oh, what a perfectly lovely mummy! Just fancy!—the poor thing—dead how many years? Something like five thousand, isn't it? And doesn't she look just like me! I mean, wouldn't she, if we had both been dead as long?"

As she said this, Miss Nitocris Marmion, the golden-haired, black-eyed daughter of one of the most celebrated mathematicians and physicists in Europe, stood herself up beside the mummy-case which her father had received that morning from Memphis.

"Look!" she continued. "I am almost the same height. Just a little taller, perhaps, but you see her hair is nearly as fair as mine. Of course, you don't know what colour her eyes are—just fancy, Dad! they have been shut for nearly five thousand years, perhaps a little more—because I think they counted by dynasties then—and yet look at the features! Just imagine me dead!"

"Just imagine yourself shutting the door on the other side, my dear Niti," said the Professor, who had risen from the chair, and was facing his daughter and the Mummy. "I don't want to banish you too unceremoniously, but I really have a lot of work to do to-night, and, as you might know, Bachelor of Science of London as you are, I have got to worry out as best I can, if I can do it at all, this problem that Hartley sent me about the Forty-seventh Proposition of the first book of Euclid."

"Oh yes," she said, going to his side and putting her hand on to his shoulder as he stood facing the Mummy; "I have reason enough to remember that. And what does Professor Hartley say about it?"

"He says, my dear Niti," said the Professor, in a voice which had something like a note of awe in it, "that when Pythagoras thought out that problem—which, of course, is not Euclid's at all—he almost saw across the horizon of the world that we live in."

"But that," she interrupted, "would be something like looking across the edge of time into eternity, and that—well, of course, that is quite impossible, even to you, Dad, or Mr Hartley. What does he mean?"

"He doesn't quite mean that, dear," replied the Professor, still staring straight at the motionless Mummy as though he half expected the lips which had not spoken for fifty centuries to answer the question that was shaping itself in his mind. "What Hartley means, dear, is this—that when Pythagoras thought out that proposition he had almost reached the border which divides the world of three dimensions from the world of four."

"Which, as our dear old friend Euclid would say, is impossible; because you know, Dad, if that were possible, everything else would be. Come, now, Annie is bringing up your whisky and soda. Put away your problems and take your night-cap, and do get to bed in something like respectable time. Don't worry your dear old head about forty-seventh propositions and fourth dimensions and mummies and that sort of thing, even if this Mummy does happen to look a bit like me. Now, good night, and remember that the night-cap *is* to be a night-cap, and when you've put it on you really must go to bed. You've been thinking a great deal too much this week. Good-night, Dad."

"Good-night, Niti, dear. Don't trouble your head about my thinking. Sufficient unto the brain are the thoughts thereof. Sometimes they are more than sufficient. Good-night. Sleep well and don't dream, if you can help it."

"And don't *you* dream, Dad, especially about that wretched proposition. Just have another pipe, and drink your whisky and go to bed. There's something in your eyes that says you want a long night's rest. Good-night now, and sleep well."

She pulled his head down and kissed him twice on his grey, thin cheek, and then, with a wave of her hand and a laughing nod towards the Mummy, vanished through the closing study door to go and dream her dreams, which were not very likely to be of mummies and fourth dimensional problems, and left her father to dream his.

Then a couple of lines from one of "B.V.'s" poems, which had been running in his head all the evening, came back to him, and he murmured half-unconsciously:

"'Was it hundreds of years ago, my love,
Was it thousands of miles away...?'"

"And why should it not be? Why should you, who were once Ma-Rimōn, priest of Amen-Ra, in the City of Memphis—you who almost stood upon the threshold of the Inmost Sanctuary of Knowledge: you who, if your footsteps had not turned aside into the way of temptation and trodden the black path of Sin, might even now be dwelling on the Shores of Everlasting Peace in the Land of Amenti—dost *thou* dare to ask such a question?" The sudden change of the pronoun seemed to him to put the Clock of Time back indefinitely.

He was standing by his desk still facing the Mummy just as his daughter had left him after saying "good-night." He was not a man to be easily astonished. Not only was he one of the best-read amateur Egyptologists in Europe, but he was also an ex-President of the Royal Society, a Member of the Psychical Research Society, and, moreover, Chairman of a recently appointed Commission on Comparative Insanity, the object of whose labours was to determine, if possible, what proportion of people outside asylums were mad or sane according to a standard which, somehow, no one had thought of inventing before—the standard of common-sense.

The voice, strangely like his daughter's and his dead wife's also, appeared to come from nowhere and yet from everywhere, and it had a faint and far-away echo in it which harmonised most marvellously with other echoes which seemed to come up out of the depths of his own soul.

Where had he heard it before? Somewhere, certainly. There was no possibility of mistaking tones which were so irresistibly familiar, and, moreover, why did they bring back to him such distinct memories of tragedies long forgotten, even by him? Why did they instantly draw before the windows of his soul a long panorama of vast cities, splendid palaces, sombre temples, and towering tombs, in which he saw all these and more with an infinitely greater vividness of form and light and colour than he had ever been able to do in his most inspired hours of dream or study?

Had the voice really come from those long-silenced lips of the Mummy of Nitocris, that daughter of the Pharaohs who had so terribly avenged her outraged love, and after whom he had named the only child of his marriage?

"It is certainly very strange," he said, going to his writing-table and taking up his pipe. "I know that voice, or at least I seem to know it, and it is very like Niti's and her mother's; but where can it have come from? Hardly from your lips, my long-dead Royal Egypt," he went on, going up to the mummy-case and peering through his spectacles into the rigid features. He put up his hand and tapped the tightly-drawn lips very gently, then turned away with a smile, saying aloud to himself: "No, no, I must have been allowing what they call my scientific imagination to play tricks with me. Perhaps I have been worrying a little too much about this confounded fourth dimension problem,—and yet the thing is exceedingly fascinating. If the hand of Science could only reach across the frontier line! If we could only see out of the world of length and breadth and thickness into that other world of these and something else, how many puzzles would be solved, how many impossibilities would become possible, and how many of the miracles which those old Egyptian adepts so seriously claimed to work would look like the merest commonplaces! Ah well, now for the realities. I suppose that's Annie with the whisky."

As he turned round the door opened, and he beheld a very strange sight, one which, to a man who had had a less stern mental training than he had had, would have been nothing less than terrifying. His daughter came in with a little silver tray on which there was a small decanter of whisky, a glass, and a syphon of soda-water.

"Annie has gone to the post, and I thought I might as well bring this myself," said Miss Nitocris, walking to the table and putting the tray down on the corner of it.

Beside her stood another figure as familiar now to his eyes as her's was, dressed and tired and jewelled in a fashion equally familiar. Save for the difference in dress, Nitocris, the daughter of Rameses, was the exact counterpart in feature, stature, and colouring of Nitocris, the daughter of Professor Marmion. In her hands she carried a slender, long-necked jar of brilliantly enamelled earthenware and a golden flagon richly chased, and glittering with jewels, and these she put down on the table in exactly the same place as the other Nitocris had put her tray on, and as she did so he heard the voice again, saying:

"Time was, is now, and ever shall be to those for whom Time has ceased to be—which is a riddle that Ma-Rimōn may even now learn, since his soul has been purified and his spirit strengthened by earnest devotion through many lives to the search for the True Knowledge."

Both voices had spoken together, the one in English and the other in the ancient tongue of Khem, yet he had heard each syllable separately and comprehended both utterances perfectly. He felt a cold grip of fear at his heart as he looked towards the mummy-case, and, as his fear had warned him, it was empty. Then he looked at his daughter, and as their eyes met, she said in the most commonplace tones:

"My dear Dad, what *is* the matter with you? If advanced people like ourselves believed in any such nonsense, I should be inclined to say that you had seen a ghost; but I suppose it's only that silly fourth dimension puzzle that's worrying you. Now, look here, you must really take your whisky and go to bed. If you go on bothering any longer about 'N to the fourth,' you will have one of your bad headaches to-morrow and won't be able to finish your address for the Institute."

She put her hand out and took up the decanter. It passed without any apparent resistance through the jar. She lifted it from the same place, and poured out the usual modicum of whisky into the glass, which was standing just where the flagon was. Then she pressed the trigger of the syphon, and the familiar hiss of the soda-water brought the Professor, as he thought, back to his senses.

But no! There could be no doubt about it. There in material form on the corner of his table was a point-blank, tangible contradiction of the universally accepted axiom that two bodies cannot occupy the same space, and that, come from somewhere or nowhere, there were two plainly material objects through which his daughter's hand, without her even knowing it, had passed as easily as it would have done through a little cloud of steam. Happily she had no idea of what he had seen and heard, and so for her sake he made a strong effort to control himself, and said as steadily as he could:

"Thank you, Niti, it is very good of you. Yes, I think I am a little tired to-night. Good-night now, and I promise you that I will be off very soon; I will just have one more pipe, and drink my whisky, and then I really will go. Good-night, little woman. We'll have a talk about the Mummy in the morning."

As soon as his daughter had closed the door, Professor Marmion returned to his writing-table. The decanter of whisky, the tumbler, and the syphon of soda-water were still standing on the corner of the table, occupying the same space as the enamelled flagon of wine and the drinking goblet which the long-dead other-self of Miss Nitocris had placed on the little silver salver.

He looked about the room anxiously, with a feeling nearer akin to physical dread than he had ever experienced before; but his worst fears were not fulfilled. Nitocris the Queen had vanished and the Mummy was back in its case, blind, rigid, and silent, as it had been for fifty centuries.

For several moments he looked at the hard, grey, fixed features of the woman who had once been Nitocris, Queen of Middle Egypt, half expecting, after what he had seen, or thought he had seen, that the soul would return, that the long-closed eyes would open again, and that the long-silent lips would speak to him. But no! For all the answer that he got he might as well have been looking upon the granite features of the Sphinx itself. He turned away again towards the table, and murmured:

"Ah well! I suppose it was only an hallucination, after all. One of these strange pranks that the over-strained intellect sometimes plays with us. Perhaps I have been thinking too much lately. And now I really think I had better follow Niti's advice, and take my night-cap and go to bed."

But as he put out his hand to take the whisky decanter he stopped and pulled it back.

"What on earth is the matter with me?" he said, putting his hand to his head. "That decanter is mine—it is the same, and yet it is standing in just the same place as that other thing—and I remember that, too. Look here, Franklin Marmion, my friend, if you were not a rather over-worked man I should think you had had a good deal too much to drink. Two bodies *cannot* occupy the same space. It is ridiculous, impossible!"

As he said the last word, his voice rose a little, and, as it seemed, an echo came back from one of the corners of the room:

"Impossible, impossible?"

There seemed to be a sarcastic note of interrogation after the last word.

"Eh? What was that?" and he looked round at the mummy-case. Her long-dead Majesty was still reclining in it, silent and impassive.

"Oh, this won't do at all! Hartley and the fourth dimension be hanged! It strikes me that this way madness lies if you only go far enough. I'll have that night-cap at once and go to bed."

He put out his hand, took hold of the whisky decanter, and as he drew back his arm he saw that instead he held the enamelled flagon in his grasp.

"Well, well," he said, looking at it half-angrily, "if it is to be, it must be."

He put out his left hand and took hold of the goblet, tilted the flagon, and out of the curved lip there fell a thin stream of wine, which glittered with a pale ruby radiance in the light of the electric cluster that hung above his writing-desk. He set the flagon down, and as he raised the goblet to his lips, he heard his own voice saying in the ancient language of Khem:

"As was, and is, and ever shall be; ever, yet never—never, yet ever. Nitocris the Queen, in the name of Nebzec I greet thee! From thy hands I take the gift of the Perfect Knowledge!"

As he drained the goblet he turned towards the mummy-case. It might have been fancy, it might have been the effect of that miraculous old wine of Cos which, if he had really drunk it, must now be more than thirty centuries old: it might have been the result of the hard thinking that he had been doing now for several days and half-nights; but he certainly thought that the Queen's head suddenly became endowed with life, that the eyes opened, and the grey of the parchment skin softened into a delicate olive tinge with a faint rosy blush showing through it. The brown, shrivelled lips seemed to fill out, grow red, and smile. The eyelids lifted, and the eyes of the Nitocris of old looked down on him for a moment. He shook his head and looked, and there was the Mummy just as it had been when he opened the case.

"Really, this is strange, almost to the point of bewilderment," he went on. "I wonder if there is any more of that wine left?"

He took up the flagon and poured out another goblet, filled and drank it.

"Yes," he continued, speaking as though under some strange exultation of the mind rather than of the senses, "yes, that is the wine of Cos. I drank it. I, Ma-Rimōn, the priest-student of the Higher Mysteries; I, whose feet faltered on the threshold of the Place of the Elect, and whose heart failed him at the portal of the Sanctuary, even though Amen-Ra was beckoning me to cross it."

"Good heavens, what nonsense I am talking! Whatever there was in that wine or wherever it came from, I think it is quite time I was off, not to old Egypt, but the Land of Nod. It seems to—no, it has not got into my head; in fact I am beginning to see that, after all, Hartley might very possibly be right about that forty-seventh proposition. Well, I will do as the Russians say, take my thoughts to bed with me, since the morning is wiser than the evening. It is all very mysterious. I certainly hope that Annie won't find these things here in the morning when she comes to clear up. I wonder what the Museum would give me for them if they were not, as I think they are, the unsubstantial fabric of a vision?"

When he got into his room and turned the electric light on, he stood under the cluster and held up his closed hand so that the light fell upon a curiously engraved scarab set in a heavy gold ring which had been given to him on his last birthday by Lord Lester Leighton, a wealthy and accomplished young nobleman who had devoted his learned leisure to Egyptian exploration and research. It was he who had sent the Mummy of Queen Nitocris to the house on Wimbledon Common instead of adding it to his own collection—not altogether unselfishly, it must be confessed, for he was very much in love with the other Nitocris who was still in the flesh.

"Now," he said, fingering the scarab, "if I was not dreaming, and if by some mysterious means Her Highness's promise is to be actually fulfilled, I ought to be able to take this ring off without opening my hand. Certainly, any fourth dimensional being could do it."

As he spoke he pulled at the setting of the scarab—and, to his amazement, the ring came off whole. There was no scar on his finger—no break in the ring.

"Good heavens!" he exclaimed, staring with something like fear in his eyes, first at his hand, and then at the ring. "Then it *is* true!" He was silent for a full minute; then he put the ring down on the dressing-table and whispered: "What a terrible power—and what an awful responsibility! Well, thank God, I am a fairly honest man!"

As he undressed he was conscious of a curious sense of reminiscence which he had never experienced before. His brain was not only perfectly clear, but almost abnormally active, and yet the current of his thoughts appeared to be turned backward instead of forward. The things of his own life, the life that he was then living, seemed to drift behind him. The facts which he had learned in his long and minute study of Egyptian history came up in his mind, no longer as facts learned from books and monuments, wall-paintings, and hieroglyphics, but as living entities. He seemed to know, not by memory, but of immediate knowledge. It was the difference between the reading of the story, say, of a battle, and actually taking part in it. He got into bed, and turned over on his right side, saying:

"Well, this is all very extraordinary. I wonder what it all means? Thank goodness, I am sleepy enough, and sleep is the best of all medicines. I should not wonder if I were to dream of Memphis again to-night. A wonderfully beautiful mummy that, quite unique—and Nitocris, too. Good-night, Nitocris, my royal mistress that might have been! Good-night!"

CHAPTER II
BACK TO THE PAST

The City of a Hundred Kings, vast and sombre, stretched away into the dim, soft distance of the moonlit night to right and left and far behind him. In front lay the broad, smooth, silver-gleaming Nile, then approaching its full flood-time, and looking like a wide, shining road out of the shadows through the light and into the shadows again—symbol of the visible present coming invisibly out of the domains of the past, and fading away into the still more hazy domain of the unknown future. Symbol, too, in its countless ripples under the fresh north wind, of the generations of Man drifting endlessly down the Stream of Time.

He was standing in the dark shadows of a huge pylon at one end of the broad white terrace of the palace of Pepi in Memphis—he, Ma-Rimōn, Priest of Amen-Ra and Initiate of the Higher Mysteries.

Nitocris was standing beside him with her hands clasped behind her and her head slightly thrown back, and as she gazed out over the river the moonlight fell full on the white loveliness of her face and into the dark depths of her eyes, where it seemed to lose itself in the dusk that lay deep down in them, a dusk like the shadow of a soul in sorrow.

He looked upon her face, and saw in it a beauty and a mystery deeper even than the beauty and the mystery of the Egyptian night as it was in those old days—the face of a fair woman, a riddle of the gods which men might go mad in seeking to read aright, and yet never learn the true meaning of it.

The silence between them had been long and yet so solemn in its wordless meaning that he had not dared to break it. Then at length she spoke, moving only her lips, her body still motionless and her eyes still gazing at the stars, or into the depths beyond them.

"Can it be true, Ma-Rimōn? Can the gods indeed have permitted such a thing to be? Can the All-Father have given His Chief Minister to be the instrument of such a foul crime and monstrous impiety as this?"

And he replied, slowly and sadly:

"Yes, it is true, Nitocris, true that thou art now Queen in the land by the will of the great Rameses; and true also it is that the shade of Nefer is now waiting in the halls of Amenti till his murderers shall be sent by the hand of a just vengeance into the presence of the Divine Assessors."

"Ah yes, vengeance," she replied, turning towards him with a gasp in her voice, "that must come; but whose hand shall cast the spear or draw the bow? We claim kinship with the gods, but we are not the gods, and what mortal hand could avenge a crime like this?"

"A woman's hand is soft and a woman's lips are sweet, yet what so cruel or so merciless in all the world as a woman? As there is nothing liker Heaven than a woman's love, so there is nothing liker Hell than a woman's hate. So saith the Ancient Wisdom, O Nitocris; and therefore, as thou hast loved Nefer the Prince, so shalt thou also hate Menkau-Ra and Anemen-Ha, his murderers and the destroyers of his promised happiness."

She shivered as he spoke, not with cold, for the breath of that perfect night was well nigh as soft as her touch and as warm as her own breath. She turned swiftly and laid her hand on his shoulder. Her touch was as light as the falling of the rose-leaves in the gardens of Sais, yet he trembled under it, and his face, which had been as pale as her own before, flushed darkly red as she looked into his eyes.

"You—yes, you, Ma-Rimōn, you too love me, do you not—truly? The stars are the eyes of the gods: they are looking on you. Tell me, do you love me? Does your blood throb in your veins when I touch you? Does your heart beat quicker when you come near me? Are your ears keener for my voice than for that of any other woman—tell me?"

His hands went up and clasped hers as they lay on his shoulders. He took her right hand and pressed it to his heart, and laid her left hand on his cheek. Then he let them fall. He stepped back, bowed his head, and said:

"The Queen is answered!"

"Not the Queen, but the woman, Ma-Rimōn, and as a woman loves to be answered. And now the woman shall speak. Nefer is dead, yet is not Nefer re-incarnated in another form, another man of another build, but yet Nefer that was—and is beside me now?"

She whispered these words very softly and very distinctly, and as the words came rippling out from between her half-smiling lips, she took half a pace forward and looked up into his face.

"Not dead—Nefer—I!" he exclaimed, starting back. "Have not the Paraschites done their work on his body? Is not his mummy even now resting in the City of the Dead? How can it be? Surely, Nitocris, thou art dreaming."

"And hast thou, a priest and sage, standing on the threshold of the Holy Mysteries, hast thou not learned the law which tells thee how, with the permission of the Divine Assessors, the souls of the dead may come back from the halls of Amenti to do their bidding in other mortal shapes? And what if they should have ordained that his soul should have thus returned?

"Thou, who art so like him that while he was yet alive mortal eyes could scarce distinguish the one from the other. May it not be that the gods, who foresee all things, made thee in the same image, perchance to this very end?"

"No, the riddle is too deep for me, even as that other riddle which I read in thy eyes, O Queen!"

"Let thy love help thee to read it, then!" she replied, coming to him and putting her hands on his shoulders again. "Tell me now, Ma-Rimōn, what wouldst thou do if thy soul were now waiting in the land of Aalu and the soul of Nefer was listening to me with thine ears, and looking at me with thine eyes?"

"And if thou——"

"Yes, and if I too believed that this were so?"

He saw the sweet, red, smiling lips coming nearer to him, and felt the soft breath on his bare throat. He saw the deep eyes melting into tenderness as the moonlight shone upon them, and in the pale olive cheeks a faint flush swiftly deepened.

"Nefer or Ma-Rimōn, I am mortal," he said, swiftly catching her wrists and drawing her towards him. "I am flesh and blood. I am man, and thou art woman—and I love thee! I love thee! Ah, how sweet thy kisses are! Now let the gods bless or curse, for never could they take away what thou hast given—and for it I will give thee all. All that has been, and is, and might have been! Priest and sage, Initiate of the Mysteries, what are they to me now! O Nitocris, my queen and my love! Sooner would I live through one year of bliss with thee than an eternity in the Peace of the Gods itself!"

The words of blasphemy came hot and fast between his kisses, and she heard them unresisting in his arms, giving him back kiss for kiss, and looking into his eyes under the dark lashes which half-hid hers; and so Ma-Rimōn, the youthful Initiate of the Holy Mysteries, became in that moment a man, and so he began to learn the long lesson which teaches to what heights and depths a woman who has loved and hated can rise and fall for the sake of her love and her hate.

"And now, my Nefer," she went on, throwing her clinging arms round his neck again, "now, good-night! Go and dream of me as I will dream of thee, and remember that, though mortals may plan, the gods decide. We may try to paint the picture, but the outline is drawn by their hands and may not be changed by ours. But, so far as this matter is concerned, I swear by the Veil of Isis, by these sacred kisses of ours, and by the Uraeus Crown of the Three Kingdoms, that, rather than be sold as a priceless chattel to grace the triumph of Menkau-Ra, I will give myself, as others did in the old days, to be the bride of Father Nile. Remember that, and remember, too, that, whatever the outward seeming of things may be, I am thine and thou art mine, as it was, and is, and shall be, until the Peace of all Things shall come."

Then the dream-vision changed from moonlight to sunlight, from night to morning; for it was the dawn of the day that was to see, as all men believed, the gorgeous ceremony of the nuptials of the daughter of Rameses with Menkau-Ra, the Mohar, chief of the House of War and mightiest of all the warriors of the Land of Khem, now that Rameses had passed from the black banks of the Nile to the shores of Amenti, and his mummy was waiting the summons of the High Gods which should recall it to life in the fulness of time and the dawn of the Everlasting Peace.

Never had even the Land of Khem seen a fairer dawn. The East shone in silver, blushed into amethyst, and flamed in gold as the Restorer of all things rose bright and glorious in sudden splendour over the City of the White Wall. Standing on the flat roof of the temple of Ptah, he looked about him in the first flush of this morning which had just dawned, big with fate, not only for him and his beloved, but also for the Land of Khem, and perchance for the world.

The great river was spreading its annual blessings over the land. The waters were broadening out into wide shining sheets, and the slow, soft music of their rippling was stealing along the great water-walls of the temples and palaces which formed the river-front of Memphis. Only a week ago the victorious armies of Khem had brought their spoils and their prisoners across the eastern frontier. There had been fruit, bread, and flesh, and wine for the poor, and banquets of royal lavishness for those who could claim right of entry into the sacred circle which enclosed the Throne, the Temple, and the camp of the victorious warrior.

For days he had heard the name of Menkau-Ra the Conqueror shouted up to the heavens by the crowds that had thronged the streets and the market-places, and, mingled with it, he had also heard the name of the girl-queen whose arms had been about his neck, and whose lips he had kissed the night before, and he knew that even now the people were asking why the Conqueror should not wed the daughter of Rameses, and become the father of a line of even greater and yet mightier Pharaohs.

He had heard their cries calmly and without anger, for he knew that that one stolen hour of sweet intercourse with her meant much more than the Conqueror himself could win—something that could not be taken by force, or even through the will of the dead king. Her soul was his, and he knew well that the man to whom she had not given her soul would never be permitted to lay a loving hand on her body.

"Ah yes, there he comes, I suppose," he went on, still talking aloud to himself, as a shrill musical peal of silver trumpets broke out from the direction of the barracks to the north of the palace. "Alas! were I but truly Nefer! That golden-crowned murderer—for sure I am that he killed him—he would not now be making ready for his triumph at the head of his victorious troops through the streets and squares of Memphis. If that were so, how glad a day this would be for Egypt and for us!"

But, as the Divine Assessors willed it, there was no triumph that day in Memphis. The sun had hardly risen to a level with the topmost wall of the Rameseum before messengers were sent out from the palace bearing the tidings that Nitocris the Queen had been stricken with a sudden malady, and that all festivities were to be deferred till the next day at the earliest.

That night, when the moon was sinking low down in the west towards the dark hills of the Libyan Desert, and the Isis Star was glowing palely like an expiring lamp hung high above the brightening eastern earth-line, he saw her muffled form gliding ghost-like towards him as he stood waiting for her on the terrace. She was clad like the meanest of her serving-maids, just as a common slave-wench who had stolen out to meet a lover of her own sort might have been. When she came within a pace of him, he held his arms out. She put hers out too, and for a moment they looked in silence into each other's eyes, and then she, seeing that the kiss which she expected did not come, parted her lips and said smilingly:

"You need not fear to kiss them, dearest, they have not yet been polluted by the lips of Menkau-Ra, although all the city has been hailing him as the betrothed of Nitocris."

Then he smiled too, and their lips met in such a long, silent kiss as only lovers give and take.

"Thy words are almost as sweet as thy kisses are, O Nitocris!" he said, "for I would sooner see thee—yes, I would sooner see thee in the hands of the Paraschites—this lovely body of thine dead—knowing that thy soul was waiting for mine on the shores of Amenti, than I would know that those sweet lips had been defiled by the touch of such as he; and yet surely thou hast spoken with him. Did he not claim the fulfilment of the promise of the great king?"

"Ah yes," she replied softly, as she slipped out of his arms, "but it is one thing to claim and another to get. Yes, I have spoken with him. I have promised all, and given nothing. I have not even yielded my hand to his lips, for I told him in answer to all the entreaties of his love—and of a truth I tell thee that he loves me very dearly, for that great, strong frame of his shook like a bulrush in the wind under the breath of my lightest words—that, until the last vows had made us man and wife, I would be his queen and he should be my subject and my slave, even as he was of the great Rameses; and with this he was fain to be content, thinking, no doubt, how soon he would be my lord and master, and I his—his queen and plaything, bound by the law that may not be broken, to submit to every varying whim and humour of his passion."

"Thy master, Nitocris! Thine! Such shame could never be. Rather would the High Gods permit Death to be the Master of Life, or Night to be Lord of Day. Is there no other way?"

"Yes, there is another way, and only one to save me, Nefer—if truly the soul of my beloved is looking out of thine eyes into mine," she whispered, coming close to him and laying her hands lightly upon his shoulders, "there is another way, but it is the way that leads through the mystery of the things that are into the deeper mystery of the things that are to be—the way of death and vengeance. Tell me, my beloved, hast thou the courage to tread it with me?"

The lovely face, the pleading lips, the searching eyes were close to his. He could feel the soft contact of her body, even her fluttering heartbeats answering his. It was the moment of the supreme test, the parting of the ways—to the heights whose pinnacles reach to the heaven of Perfect Knowledge, or to the abysses whose lowest depths are the roof of hell; for there is but one heaven and one hell, and their names are Knowledge and Ignorance.

There lay the fulfilment of his vows, the renunciation of the lower life with all its potent witcheries of the senses, with all its exquisite delights and glittering prizes, fame and honours, power and wealth, and, dearest of all, the love of woman.

Here, clasped in his arms, stood Nitocris, her hands still resting lightly on his shoulders, her head lying on his breast, her eyes upturned, the star-beams swimming in their luminous depths.

"Nefer, beloved, answer me!"

The stars grew dim, and the solid floor of the terrace shook under his feet. He bent his head and laid his lips upon hers.

"Thou art answered, O Nitocris—even unto death and the life beyond!"

Her lips returned his kisses—kisses that were curses—and then for many minutes they conversed in hurried whispers. At last she slipped out of his arms and left him, his lips burning from the clinging touch of hers, and his heart cold with a fear that was greater than the fear of death.

He clasped his hands to his temples and looked up at the coldly shining Isis Star, and through the silence there came to his soul in the speech that is never heard by the ears of flesh the fateful words:

"Once only is it given to mortals to look into the eyes of Isis. He who looks and turns his gaze aside has found and lost."

CHAPTER III
THE DEATH-BRIDAL OF NITOCRIS

The day of the bridal of Nitocris the Queen with Menkau-Ra the Conqueror had come and gone in a blaze of golden splendour. In all the Upper and Lower Lands no head was held so proudly as the head of Menkau-Ra, no heart beat so high as his that day, nor did any cheek bloom so sweetly, or any eyes shine so brightly as the cheeks and the eyes of Nitocris—so strange are the workings of a woman's heart, and so far are its mysteries past finding out.

And now the bridal feast was spread in the great banqueting hall which Pepi the Wise had made deep down in the foundations of his palace below the waters of the Nile at flood-time, and at midnight the waters would be at the full. It was here that Nitocris had sat at the betrothal feast with Nefer but a few hours before his death, for here he had drunk from the poisoned cup which Anemen-Ha the High Priest had prepared, and here only would Nitocris meet her guests.

The great hall shone with the light of a thousand golden lamps, which shed their radiance and the perfume from the scented oils in which were dissolved the most precious gums of the distant East.

The long tables, spread with snowy linen and loaded with vessels of gold and silver and glass of many hues and curious forms, flashed and glittered in the glow of the thousand flames. The vineyards of Cos and Sais had yielded their oldest and sweetest wines, red and purple and golden. The choicest meats and the rarest fruits that ripened under the glowing suns of Khem—all was there that could make glad the heart of man and fill his soul with contentment.

At the centre of the table, which stood on a raised platform in front of the great black pedestal of the Colossus of Pepi, Nitocris the Queen sat in her chair of ivory and gold, clad in almost transparent robes of the finest silk of Cos, shining with gems, and crowned with the Uraeus Snake, and the double diadem of the Two Lands.

On her right sat Menkau-Ra, crowned and robed in royal vesture, and on her left Anemen-Ha in his priestly garments of snowy linen. At the other tables sat their friends and kindred, the families of the Mohar and the High Priest, the chief officers of the victorious army and all the proud hierarchy of the Temple of Ptah, for was not this the triumph of Anemen-Ha no less than of Menkau-Ra?

Only Ma-Rimōn was absent. He had disappeared from the temple early in the morning, and no one had given a thought to his going, for one base-born, even though of royal blood, had no place at the bridal feast of the Queen and her chosen consort.

The libations had been poured out to the Lords and Ladies of Heaven—to Ptah the Beginner, and Ra the Lord of Day, to Sechet the Lady of Love and War, and Necheb the Bringer of Victory; and when the slaves had carried round the viands till all were satisfied, the guests were crowned with garlands, and the jars of the oldest and choicest wines were broached. The feast was ended, and the revel was about to begin.

The last half of the last hour of the night was well-nigh spent, and while the guests were waiting for the signal from the royal table, the Queen rose in her place, and, in the silence that greeted her, her voice sounded sweetly as she spoke and said:

"O my guests—ye who are the holiest and the bravest in the Land of Khem, though our hearts are joyful, and our souls refreshed with wine and good cheer, let us not forget the pious customs and wise ways of our ancestors, for it is fitting that in such hours as this our hearts should be turned from pride by the remembrance that we live ever in the presence of death, and that this world is but the threshold of the next. Ill, too, would it become me to forget, in the midst of my present happiness, to pay the honour due to him who might have shared this crown with me; wherefore let the noble dead be brought into our midst, so that the soul of Nefer, looking down from the flowery fields of Aalu, may see that in the hour of our joy we do not forget the sorrow of his untimely death."

Then she clapped her hands, and Menkau-Ra and Anemen-Ha shifted in their seats, and looked at each other with eyes of evil meaning as six slaves appeared at the lower end of the hall, bearing upon their shoulders the mummy-case of Nefer, the dead Prince, beloved of Nitocris. Now low, sad music sounded from a hidden source, and to the cadence of this the slaves marched slowly round the tables, followed by the eyes of the silenced and sobered guests. Then they stopped in front of the Queen's seat, and she said:

"Let the case be set up against the central pillar yonder, and let the face of the Prince be uncovered, that I may look upon him who was to have been my lord."

"But if I may speak, Royal Egypt," said Anemen-Ha, the chief of the House of Ptah, leaning towards her, "that would be beyond the law of the gods and the customs of the land. To look on the face of the dead were defilement for thee and us."

"Yet this once it shall be done, O Priest of the Father of the Gods," answered Nitocris, turning and looking into his eyes, "for last night I had a vision, and I saw the soul of Nefer come back to his mummy, here in this hall, at my bridal feast, and his eyes opened, and his lips spoke, and made plain to me many things that I greatly longed to know. But why shouldst thou turn pale and tremble, thou the holiest man in the land? What hast thou to fear, even if my vision came true? And thou, too, Menkau-Ra the Mighty, hast thou slain thy thousands, and yet fearest to look upon the face of one dead man? See, see!" and she pointed her finger at the face of the mummy. "By the power of the just and merciful gods, my vision shall be made very truth indeed! Look, Anemen-Ha, Priest of the God who is King of Gods! Look, Menkau-Ra, thou who wouldst reign in the place of Nefer. Behold, he has come back from the bosom of Osiris to greet thee!"

With eyes fixed and ears sharpened by such terror as only the sin-steeped soul can know, they saw the waxen eyelids of the mummy slowly rise, the dim, glazed eyes look out from underneath them, the dry, black lips move, and heard a thin, harsh voice say through the awful silence:

"Greeting, Nitocris, my Queen—greeting from the gloom of Amenthes, where I have waited too long for those who ere now should have stood with me in the Halls of Doom and the presence of the Assessors! Say now, thou who sittest feasting between my murderers, how much longer must I wait for thee and them?"

Not long, O Nefer, my beloved, not long! Tarry yet a little while, O outraged soul, in the shape that once was thine, and thou shalt see thyself avenged. Lo, I hear the wings of Kefa, Goddess of the Flood-time, rustling in the silence of the midnight skies. She herself shall pour out a libation to thine injured shade! "Nay, nay, my lords, and you good friends of those who did my own true lord to death, sit still, and drain a farewell cup with me, your Queen. It is too late to fly, for every way is closed. The High Gods have spoken, and I will do their bidding!" Then, extending her white, jewelled arms toward the mummy, she cried in a deeper, harsher tone: "O Nefer, my Prince and my love! There lives no man in Khem who shall take thy place beside me, or usurp the throne that should have been thine. I have sinned, but I repent me of the wrong. Lo, now I come and bring thee a goodly sacrifice to cheer thine angry heart—my lord, my love, I come!"

Held by the triple spell of guilt and fear and wonder, they listened to these terrible words in silence, white horror sitting on their blanching cheeks and brows.

As she ceased she raised her arms above her head, a golden cup full-crowned between her glittering hands. A moment she held it aloft, then dashed it to the floor, and cried in a voice that rang like the laughter of devils through the awful silence:

"Come, Kefa, come, and bear me to my lord!"

The goddess answered in a mighty rush and roar of waters, long pent and swiftly loosed. Then above the tumult rose the hoarse shouts of men and the shrill screams of women, and the crash and clash of tables overturned; then came the swirl and bubbling hiss of a flood that gleamed darkly under the golden lamps and swiftly rose towards them, bearing upon its surface white arms with outstretched hands gripping at the empty air, and gauzy robes which half hid gleaming limbs, white faces with wildly-staring eyes, and teeth that grinned between tight-drawn lips so lately smiling; strong swimmers fighting for another moment's breath, and one by one dragged down by many hidden hands: then the sharp hiss of swift-quenched flames, then darkness, and the stifling of sobbing groans into silence, and after that only the sibilant undertone of waters rushing swiftly past smooth walls through utter night.

"Dear me!" the Professor heard himself say as he sat up and rubbed his eyes, "what on earth can be the matter with me? Egypt—the Queen—Palace of Pepi—bridal feast of Nitocris and Menkau-Ra—yes, yes, of course I remember it all now. She made me impersonate Nefer in the mummy-case, and then, when she had frightened her guests half out of their wits, she avenged her lover by opening the sluice-gates and drowning the lot, herself included. A rare device, that of old Pepi's, for getting rid of hospitably entertained enemies. Not quite in accordance with our modern ideas of sport, I'm afraid, but in those days we thought a good deal more of effectiveness than sport. Good heavens! What sort of nonsense am I talking? Dreaming, I suppose."

He stopped as the reflection of a brilliant flash of lightning lit up his window, and bursts of rain dashed upon the panes.

"Ah yes, of course, that's it! Quite in accordance with the theory of dreams. It's only the difference between a thunder-shower and the Nile flood. The Genius of Dreams could easily account for the rest. Certainly this apparatus that we call our brain plays some very curious tricks with us sometimes. I suppose this is one of them. And yet if ever there was a dream that seemed like reality that one did. The Mummy and the long-dead Nitocris back to life! By the way, I wonder whether that flagon was really there, and whether there *was* any wine in it? If there was, perhaps I took too much of it. Ah, there's the rain again!

"By the way now, suppose that this fourth dimension that has puzzled so many of us is, after all, duration? If so, it would solve a great many problems, because it would be possible to be and not to be at the same time, and, therefore, for two bodies to occupy the same space. That would be perfectly easy of supposition to the being to whom time and eternity were one. Yes, I believe that when the great problem is solved, it will be found that the fourth dimension *is* duration, extending in all directions like the circumference of a circle, the edges of a cube, and the curves of the conic sections.

"Yes, I really do think I have got it at last, and that confounded Mummy has taught it me. Still, I don't think I ought to speak as disrespectfully as that of a young lady who has been dead for the last fifty centuries or so and has come back. Yes, that is it. It *is* duration."

Perfectly satisfied for the time being with this solution, he turned over on to his right side—for, to his disgust, he found that he had been lying on his back, a most pernicious position where dreaming is concerned—and went to sleep. Half an hour later he was awakened by another heaven-shaking crash of thunder.

CHAPTER IV
THIEVES IN THE NIGHT

This time he was very much awake. In fact, his sense of wakefulness seemed almost superhuman. His faculties were preternaturally alert, and he had a feeling of what might properly be called mental extension—it was not exaltation—- which seemed to widen his mental vision enormously. Problems which had puzzled him to desperation suddenly became as obvious as the first axioms of geometry. In short, he felt as though he had become a new man, re-born, or re-incarnated, into another world which contained the one he had so far lived in, but which was infinitely vaster in some undefined way which was not yet plain to him.

He lay for some time thinking over the extraordinary happenings of the evening and his dream, which he remembered with astonishing exactness of detail. Then a sudden turn of thought carried his mind to the subject of miracles, apparitions, ghosts, and mathematical impossibilities such as squaring the circle and doubling the cube—and to his amazement he found that the impossible of yesterday had become the possible—nay, the almost absurdly obvious of to-night.

He went on thinking and wondering until he began to half-believe that he was dreaming again, so he got up and switched on the electric light. Then he turned involuntarily towards the wardrobe, which, as usual, had a long mirror running down the middle of it. To his amazement he did not see himself reflected in it. The mirror seemed to have vanished, and in its place was a window looking into his study.

He saw the mummy-case leaning up against the wall, but it was empty. In front of it stood a man and a woman. Both were plainly, almost meanly, dressed; the man in a tightly-buttoned black frock-coat and baggy grey trousers; the woman in a plain gown of dark stuff, and a shawl which was draped round her head and shoulders in somewhat Eastern fashion.

He could see their faces distinctly in profile. They were of the classic Coptic type which so persistently reproduces the features of the old Egyptians as we see them outlined in the wall-paintings of the temples and the half-mutilated carvings and statues. The window of the study was open, but the door was shut; so was the door of his own room, but for all that he distinctly heard the man say to the woman in Coptic, which, curiously enough, sounded as familiar to his ears as the faces seemed to his eyes:

"Neb-Anat, it is gone! These heathen ravishers have not been content with stealing the body of our Queen from its sacred resting-place and bringing it here, whither we have traced it with so much labour. See, it has been stolen again; hidden, no doubt, so that the servants of the King could not find it. It may be that even we have been suspected and watched, in spite of all our care. Yet it must be found, or the doom that may not be revoked will be ours."

"Even so, Pent-Ah," replied the woman in a soft, musical voice which well suited the comeliness of her face; "but though the priceless treasure has been taken from its casket, it cannot have been carried out of the house, for you know that every approach has been watched closely since it was brought here. Come, in this house it must be, and to find it is our task. Every one is asleep; take off thy shoes and let us search."

She took off her own shoes as she spoke, and he saw the man do the same. Then, as the man opened the door and they passed out of the study, the picture vanished from the mirror.

Amazement at what he had seen and heard—the disappearance of the Mummy, the presence of the man and woman, evidently charged with what they believed to be the sacred mission of stealing it back again, and their evident purpose of searching the house for it—instantly gave place to a quick thrill of fear.

His daughter's bedroom was on the same floor as the study, only a couple of doors away round the corner of the landing. These people would search every room. What if she had not locked her door securely, or if they had some means of opening it? She was the living image of the dead Nitocris. He did not dare to think of what might happen to her. Would these new-found, strangely-given powers of his suffice to protect her? If not, he would have but little use for them, since she was his nearest and dearest on earth.

He pulled his stockings over the pants of his pyjamas and put on his velvet working jacket, forgetting for the moment that, if these things were true, it would be perfectly easy for him to make himself invisible to beings in the ordinary world of three dimensions. Then he turned out the light, opened the door very softly, and crept downstairs.

Yes, what he had seen was true. He heard the soft, shuffling patter of stockinged feet along the landing, though he could see nothing in the dark. A door opened gently. His sense of location told him that it was the door of the spare bedroom next but one to the study. He felt his way silently and softly along the wall, and as he did so his hand touched the electric switch. Should he turn the light on and alarm the house? Whoever was there had "broken and entered" after midnight, and was therefore outside the law. No, he would not do that. If what he had seen was true, the intruders believed that their mission was a sacred one. No doubt the man was armed, and perhaps the woman also, and what would a knife-stab mean to them on such a desperate quest?

As these thoughts ran at lightning speed through his mind, he saw a faint glow inside the room. He crept forward and looked round the side of the doorway. The man had a little electric lamp in his hand and was flashing the slender rays all over the room. He drew his head back quickly as he heard him say:

"There is nothing here, Anat. Come, let us try the next room. Neither lock nor bolt nor even human life must stand in the way of our search now that we have begun it!"

He heard them coming towards the door. Instinctively he shrank back, and his heart stood still as he thought of what would happen if the man chanced to turn the little ray of his lamp on him. Almost involuntarily his thoughts went back to the promise of Queen Nitocris, and something like a prayer that it might be kept rose to his lips.

They came out, and the man flashed the thin electric ray up and down the passage. It wavered hither and thither, and at last fell directly on his face. He was anything but a coward, but he was thinking of Niti—and what if a knife-stab left her undefended? But to his amazement, although they were both looking straight at him, the expression of neither face changed in the slightest. They had not seen him. The Queen had answered his prayer. He was no longer in the world of three dimensions, and so he was invisible to all dwellers in it. For him, then, there was evidently no danger—but Niti——?

They moved along to the next door. That was hers. The woman put her hand on the knob and turned it. To his horror, the door opened. She had forgotten to lock it. They both crept in, and he followed them boldly enough now, knowing what he did. The ray leapt rapidly about the room till it fell on the bed with its pale blue silken coverlet, and then on the pillow, on which rested the head of the sleeping, breathing image of the long-dead Queen. ·

With a half-stifled gasp the man shrank back and dropped the lamp, and the Professor heard him say to the woman in a shuddering whisper:

"By the High Gods, Neb-Anat, it is a miracle! Do you not see her? It is she—the Queen—alive again, as the ancient prophecy said she should be. What magic have these heathens used?"

"Yes," replied the woman, whispering lower, "truly it is the Queen, and she is alive and sleeping—no doubt passing from the sleep of death through the sleep of life to life again. Now, O Pent-Ah, is our task much harder, yet will its accomplishment be all the more glorious for you and me, and greatly will our Lord reward us if we can restore to his keeping, not the ravished mummy of Nitocris, but the Queen herself, warm and breathing and beautiful, as she was in the ancient days of the great Rameses."

"I'll be hanged if you do!" said the Professor to himself, "not, at least, if Her Majesty's legacy to me is worth anything. Abduct my daughter at the dead of night, would you, you scoundrels? We'll see about that. If you don't leave this house as thoroughly frightened as ever you were in your lives, I know nothing about the fourth dimension."

Meanwhile he heard them both groping about the floor after the lamp. The woman found it, and pressed the button. The ray fell on the man's face, and he saw that the olive of his skin had turned to a ghastly grey. His eyes were wide open, and his mouth and nostrils were working with intense excitement. Then the woman turned the ray on Niti's face again.

"They will wake her if this goes on much longer," said the Professor to himself again. "I had better stop this little comedy before it becomes a tragedy. Poor Niti would go half mad if she found these two scoundrels by her bedside—and yet if I do anything out of the way they will yell. Ah, I think I have it!"

He walked softly out of the room, and when he got into the passage he whispered in the tongue that had become so strangely familiar to him:

"Pent-Ah, Neb-Anat, come hither instantly! Who are you that you should disturb the slumbers of your Lady the Queen!"

He saw them stare at each other with eyes wide with fear and wonder.

"It is the command of the Mighty One," whispered the woman, taking hold of the man's hand and drawing him towards the door.

"And He must be obeyed," said he in reply, bowing his head and following her.

They closed the door very softly behind them.

The Professor could not repress a sigh of thankfulness for Niti's escape from what, at best, would have been a very terrible fright.

"And now, my friends," he went on to himself, "I think I can teach you not to come into an English gentleman's house again with an idea of stealing his property, to say nothing of abducting his daughter."

The man and woman were still staring at each other by the light of the lamp, each holding each other's trembling hand, when the lamp was suddenly snatched away from the woman and went out. Then, to their horror, the ray shot out again in front of them as though the lamp were floating by itself in the air. It flashed from face to face, both ghastly with fear. Then an invisible hand gripped the man's, and drew him with irresistible force along the passage. The woman grasped his coat, and followed with shuffling feet and shaking limbs, dumb with wonder and fear. The hand led them down the passage, round the corner, and into the study. Then it released them. They heard the door shut and the key turn in the lock. Then there was a click, and the electric cluster above the writing-table shone out, apparently of its own volition. The woman uttered a low scream, and cowered down in a corner of a big sofa that stood by the bay-window. The man, after one terrified glance round the room, began to creep towards the open sash; but the invisible hand gripped him by the collar and pulled him back. His trembling knees gave way under him, and he rolled in a heap on the floor.

Then, to his wondering horror, he saw a stout blackthorn stick which was standing in a corner of the room, jump up into the air and leap towards him. He put his head down on to the carpet, covered his eyes with his hands, and began to moan with terror. The stick came down with what seemed to him superhuman force again and again on his back and shoulders. He whimpered and moaned, and at last howled with pain. He rolled over and looked up, and there was the stick hanging in the air above him. He put up his hands clasped as though in prayer, and down it came on his knuckles. He did not howl this time. His hands unclasped and dropped beside him; his head went back, and he fainted in sheer terror.

"There, my friend," said the Professor aloud, forgetting the presence of the woman for the moment; "mummy or no mummy, I don't think you will come into this house again. And as for you, madam," he went on, "of course, I can't give you a hiding, so the sight of his punishment will have to be enough for you. Still, I think you have had enough of attempted mummy-stealing to last you some time."

The woman stared up into the vacancy out of which the voice came, her eyes dilated, and her lips trembling with the movement of her lower jaw. She saw a jug of water get up off the table and empty itself over her companion's face. Then she fainted, too.

When Pent-Ah came to himself and sat up, he saw an elderly gentleman, tall and erect as a man in the prime of life, standing over him with the blackthorn in one hand and the water-jug in the other.

"I am not going to ask what you two are doing here," he said sternly, "because I know already. If I called the police I could send you both to prison for house-breaking and attempted robbery; but I don't want any fuss, and perhaps you have been punished enough for the present. Ah, I see your accomplice is coming round. You came in by the window, I suppose. Now get out by it as quick as you can, and mind you keep your mouths shut as to what has happened to-night. If you don't," he went on, suddenly changing into Coptic, "beware of the anger of your Lord—of Him who never forgives!"

The man scrambled to his feet, whimpering:

"I go, Lord, I go, and my lips shall be silent as the lips of——"

He cast a frightened glance towards the mummy-case, and then, grasping the woman roughly by the arm, he dragged her towards the open window, saying:

"Come, Neb-Anat, come ere the wrath of our Lord consumes us!"

"Why, where's the Mummy, Dad?" said Miss Nitocris, as she came into her father's study just before breakfast the next morning, and looked in amazement at the empty case.

"Stolen, my dear, I am sorry to say," replied the Professor gravely. "Did you hear any noises in the house last night, or were you sleeping too soundly?"

"I seem to have an idea that I did," she said, "but only a dim one; I thought I only dreamt it. But did you, Dad? Do tell me all about it. What a horrible shame to steal that lovely Mummy! And it was so like me, too. I believe I should have got quite fond of it."

"Yes, dear," continued the Professor, speaking, as she thought, a little nervously. "There was a noise, and I heard it. I came down here and turned the light on. I found the window open and the Mummy gone—and that is all I can tell you about it."

CHAPTER V
ACROSS THE THRESHOLD

After breakfast Professor Marmion, according to his practice on fine days, lit his pipe, and went out for a stroll on the Common to put in a little hard thinking, while Miss Nitocris, after seeing to certain household matters, sat down in his study and read the papers, in order that she might be able to give him a synopsis of the world's news at lunch. He did not read the newspapers himself, except, perhaps, in the train, when he had nothing better to do. He took no interest in politics, for one thing, and he had still less interest in professional cricket and football, racing, and what is generally called sport. He had a fixed opinion that all the events happening in the world which really mattered, not even excepting the proceedings of learned societies and the criminal and civil Law Courts, could be adequately recorded on a couple of sheets of notepaper. In other words, he had an absolute contempt for everything that makes a newspaper sell, and therefore his daughter had very soon learnt to omit these fascinating items entirely.

Curiously enough, his mind seemed to be running on this subject of all things that morning. He had been reading an article in the *Fortnightly* on the growing sensationalism, and therefore the general decadence of the English Press a day or two before, and this had got connected up in his thoughts with the amazing happenings of the last twelve hours, and he asked himself what would happen if he were to give the narrative of his experiences in a letter to the *Times*, supported by the authority of his own distinguished and irreproachable name.

Certainly it would be the most sensational communication that had ever appeared in a newspaper. In a day or two, granted always that the *Times* had no doubts as to his sanity and printed the letter, the whole Press would be ablaze with it; Wimbledon would be besieged by reporters eager to see miracles; and then they would go away and write lurid articles, some about the miracles, if they saw them, and some about an absolutely new form of conjuring that he had invented. Then the scientific Press would take it up, and a very merry battle of wits would begin. He smiled gravely as he thought of the inkshed that would come to pass in a *combat à l'outrance* between the Three Dimensionists and the Four Dimensionists, and how the distinguished scientists on each side would hurl their ponderous thunderbolts of wisdom against each other.

Then there would be the religious folk to deal with, for naturally no theologian of any enterprise or self-respect could see a fight like that going on without taking a hand in it. The Churches, of course, had a monopoly of miracles, or at least the traditions of them. The Christian Scientists, blatantly, claimed to work them now, but their subjects died with disgusting regularity. So he quickly came to the conclusion that, if he were once to state in plain English that he could accomplish the seemingly impossible; that he, a mere mortal, could make himself independent of the ordinary conditions of time and space and break with impunity all the laws which govern the physical universe, he would simply make himself the centre of a vortex of frenzied disputation which would shake the social, religious, and scientific worlds to their foundations, and that would certainly not be a pleasant position for an eminent and respected scientist, who was already a certain number of years past middle age—to say nothing of the very real harm that might be done.

Of course, he could settle all the disputes instantly, and dazzle the whole world into the bargain by simply delivering a lecture, say, before the Royal Society, on the existence of a world of four dimensions, and then proving by ocular demonstration that it does exist; but what would happen then? Simply intellectual anarchy.

Every belief that man had held for ages would be negatived. For instance, if there is one dogma to which humanity has clung with unanimous consistency, it is to the dogma that two and two make four. What if he were to prove—as, of course, he could do now that this mysterious hand, outstretched through the mists of the far past, had led him across the horizon which divides the two states of Existence—that, under certain circumstances, they would also make three or five? What if he demonstrated that even the axioms of Euclid could, under different conditions, be both true and false at the same time?

No, the thought of overthrowing such a venerable authority and plunging the scientific world into a hopeless state of intellectual chaos sent a shudder through his nerves. He could not do it.

And yet it was only the bare, solid truth that he did possess these powers. The dream of the death-bridal of Nitocris might possibly have been nothing more than just a dream, or possibly the revival of an episode in a past existence; but the other experiences certainly were not. He had taken off his ring without unbending his finger. Yes, he could do it again now; it was just as easy as taking it off in the ordinary way. He certainly had not been dreaming when the Mummy had become Queen Nitocris and given him the wine. He could not have been mad or dreaming, because his daughter was there. The episode of the strange stealers who had come into his house—that too was real, for they had left their lamp and the man's shoes behind them, and the Mummy was gone!

He took a piece of string out of his pocket, tied the two ends, and then with the greatest ease tied another knot in the string without undoing the first.

A motor-car came humming along the road towards him, and he began to think what this place was like a thousand years before motors were heard of. That instant the motor vanished, and he found himself standing in a little glade surrounded by huge forest trees with not so much as a foot-track in sight. He made his way through the trees in what he remembered to be the direction of the road, and presently, through an opening avenue, he saw the sun glittering upon something moving, and heard voices; and then past the end of the avenue half a dozen armoured knights, followed by their squires and a string of men-at-arms guarding a covered waggon, and after these came a motley little crowd of travellers, some on horseback and some on foot, evidently taking advantage of the escort to protect them from robbers.

"Dear me!" said the Professor to himself, not without a little shiver of apprehension, "this is very interesting. I seem to have put myself back into the tenth century. Yes, that is certainly tenth-century armour that they're wearing. I mustn't let them see me, or there's no telling what they'd think of an elderly gentleman in a soft hat and a twentieth-century morning suit. But perhaps," he went on with his reasoning, "they can't see me at all. My condition is N to the fourth now. There's a thousand years between us; I forgot that. At any rate, I'll try it."

He walked quickly down the avenue, and stood by the side of the rugged path looking at the strange spectacle. No one took the slightest notice of him. And then a chill of awful loneliness struck him. Although he could see and move and hear, and, no doubt, eat and drink in this world, he was unexistent as regards the inhabitants of it, and yet he knew perfectly well he was standing by the side of the road where the motor-car ought to be, and over there, a few hundred yards away, Niti would be sitting in her room or walking in the garden—and she wouldn't be born for nearly a thousand years yet.

It was certainly somewhat disquieting, this power of living in two existences and different ages, but it was a matter that would take some little time to get accustomed to.

The next instant the cavalcade and the forest had vanished, and there was the motor-car, just spinning past him. He was on the Wimbledon Common of the twentieth century once more. He stroked his clean-shaven chin with his finger and thumb, and walked slowly along the path by the side of the road, and then across the grass towards the flagstaff.

"I think I begin to see it now," he murmured. "Of course, life, that is to say real, intellectual, or, as some would say, spiritual life, is, after all, the coefficient of that totally unexplainable thing called thought which enables us to explain most things except itself. Time and space and location are only realities to us in so far that we can see them. A human being born blind, dumb, deaf, and without feeling would still, I suppose, be a human being, because it would be conscious of existence; it would breathe and know that its heart was beating, but without sight or sensation there could be no idea of space—time, to it, would be a meaningless series of breaths or heartbeats. Without touch or sight it could have no idea of form or size, which are merely conditions of space, and both the past and the future would be absolutely non-existent for it."

He paused, and walked on a little way in silence, arguing silently with himself as to the correctness of these premises. Then he began aloud again:

"Yes, I think that's about right. And now, suppose that such a being became endowed with the natural senses, one by one. It would go through all the processes of the physical and mental evolution of humanity until it reached the highest of human attributes—the ability to think, and therefore to reason. In other words, from a merely living organism it would, in the old Scriptural language, have become a living soul. That is, obviously, what the words in Genesis were really intended to mean. It would then become capable of development, of proceeding from the partly-known to the more fully known, until, granted perfect physical and mental health, it reached what are generally called the limits of human knowledge."

The Professor's thumb and finger went up to his chin again. He walked another two or three hundred yards in silence; then he recommenced his spoken argument with himself:

"Limits of human knowledge? Yes, that sounds all very well in ordinary language, but are there any? Who was it said that a man trying to reach those limits was like the child who saw a rainbow for the first time, and started out to find the place where it rested? The simile is not bad, not by any means. Just in the same way, we try to imagine the limits of time and space, and we can't do it. Only infinity of space and duration are possible, and yet we can't grasp them; still, they are the only possible states in which we can exist. And now, as I have had a glimpse of the past, I wonder what this place would be like in ten thousand years?

"Good heavens, how cold it is!" He shivered, and buttoned up his coat, and continued, looking about him on the vast snow-field dotted with hummocks of ice which lay bleak and lifeless about him: "Ah, I suppose either the Gulf Stream has got diverted, or the earth's axis has shifted and we are in another glacial epoch.

"We!"

Again the shock of utter isolation struck him, but it seemed to hit him harder this time. The world that he had been born in lay ten thousand years behind him. For all he knew, he might be standing upon what was now the earth's North Pole. Civilisation, as he had known it, might have been wiped off the face of the earth, and the remnants of humanity flung back into savagery. He looked up at the sun, and saw that it was almost exactly where it had been, and that it had not perceptibly diminished in power.

The idea was not at all pleasant to him, and very naturally his thoughts turned back once more to his cosy home that had been on the edge of Wimbledon Common ten thousand years ago. He remembered, with a curious sort of thrill, some notes which he had to complete that morning for his lecture—and in the same instant he was walking back across the turf towards his house through the warm May sunshine.

"Yes," he said to himself, as he drew a deep breath of the sweet spring air. "I was right; that's it. The fourth dimension is a form of duration in some way correlated with space. I shall have to work that out in the light of the greater knowledge, which Her vanished Majesty has given me, and which I almost attained to in Egypt. Wherefore, existence in a state of four dimensions, or the world of N^4, as I have always called it, is, roughly speaking, one. Time and space are, as it were, two sides of the same shield, and a person living in that world can see both of them at once. Wherefore, past, present, future, length, breadth, thickness, here and there are all the same thing to him. It's a great pity there isn't a fourth dimensional language as well, so that one could state these things a little more precisely. But that, of course, is out of the question.

"Really, I can hardly make myself understand it as far as words and phrases are concerned; still, there it is; and now the question arises: Having got this power, as I certainly have, of transferring myself from one existence to another by a mere effort of thought, because it is very evident that this power is really only an extension or an exaltation—confound the language of the third dimension—I can't say it! Although I understand what it is, it won't go into words. What am I to do with it? Its possibilities are, of course, a little appalling—that is to say, from the point of view of N^3. I have not the slightest desire to shake the fabric of Society to pieces, as I could do, and still less have I taste for spending the rest of my scientific career in what the world would very easily believe to be conjuring tricks. I hope I am not going to be another of the unnumbered proofs of Solomon's wisdom when he said, 'Whoso getteth knowledge, getteth sorrow.' I wonder what sort of advice Her late Majesty of Egypt——

"Dear me, what nonsense I am talking! Her late Majesty? That won't do at all—she has reached the Higher Plane too, so, of course, she can't be dead——"

And then with the force of a powerful electric shock, the terrible fact struck him that, for those who had reached that plane, there was no death! Here was a new light on the weird problem which he had somehow been called upon to deal with.

"I wonder what Her Majesty would really think of it?" he murmured, after a few moments of mental bewilderment. "Dear me, who's that?"

He looked up, and, to his utter amazement, he saw Queen Nitocris, arrayed exactly as she had been on that terrible night of her bridal with Menkau-Ra, walking towards him; a perfect incarnation of beauty, but——

"Oh dear me!" said the Professor, "this will never do. Good heavens! everybody in Wimbledon knows me, and—well, of course, Her Majesty is very lovely and all that; but what on earth would people think if any one saw me strolling across the Common in company with an Egyptian Queen—to say nothing of the costume—and the image of my own daughter, too!"

The figure approached, and the Queen, dazzlingly and bewilderingly beautiful, held out her hands to him, and their eyes met and they looked at each other across the gulf of fifty centuries. Impelled by an irresistible impulse coming from whence he knew not, he clasped them in his, and said, apparently by no volition of his own, in the Ancient Tongue:

"Ma-Rimōn greets Nitocris, the Queen! What hath he done that he should be once more so highly honoured?"

At that moment a carriage came by along the road quite close to them. Two of its occupants were looking straight towards them. They passed without taking the slightest notice, as they must have done had they seen such a marvellous figure as that of the Queen. And then he remembered that, unless she willed it, no one in the world of N^3 could see her, since it was for her, as it was for him now, to make herself visible or invisible as she chose to pass on to or beyond the lower Plane of Existence. These things were quickly becoming more plain to his comprehension, although, as will be readily understood, it was not a lesson to be learnt very easily.

"Welcome, Ma-Rimōn," replied the Queen, in a voice which filled him with many distant and strange memories, "but let there be no talk between us of honour, for in this state there is neither honour nor dishonour, neither ruler nor subject, neither good nor evil, since all these are absorbed in the Perfect Knowledge. Yet it is the will of the High Gods that I should help thee and guide thee in that new world whose threshold thou hast so lately crossed. It was my hand led thee from the path of Light to the path of Darkness, and for that I have paid the penalty as well as thou.

"For many ages, as time is counted in that other world, we have toiled, sometimes together, sometimes apart, sometimes in honour, sometimes in dishonour, yet ever struggling on to regain the heights which then we had so nearly won. The High Gods permitted me to reach them first, and therefore it was my hand which was stretched out to lead thee across the Border.

"Now, my message to thee is this: Thou hast powers which no other man living in that lower state possesses; see to it that they be used rightly. Forget not that in that other world sin and shame, oppression and misery, are as rife as, within the limits of time, they have ever been. Make it thy concern that the forces of evil shall be weaker and not stronger for the use of these powers to which thou hast attained.

"We shall meet often in that other world, and that living other-self of mine, thy daughter in the flesh and bearer of my name, through every moment of her time-life, I shall watch and guard her, for she, too—although she knows it not—is approaching the light never seen by the Eye of Flesh, and, though strange things should befall her, it will be for thee in that other state, knowing what thou dost in the Higher Life, to help me in this task as in others. Now, farewell, Ma-Rimōn," she said, holding out her hands again.

As he took them, they melted in his grasp, two lustrous eyes looked at him for a moment and grew dim, and he was once more alone on Wimbledon Common.

"I think I'll be getting home," he said, looking at his watch, and he turned and walked slowly with bent head and hands clasped behind his back to the house.

CHAPTER VI
THE LAW OF SELECTION

In actual mundane time, to use a somewhat halting expression, Professor Marmion's walk had occupied about a couple of hours. His strange experiences had, of course, occupied none, since they had taken place beyond the bounds of Time.

Meanwhile, Miss Nitocris had finished her digest of the morning papers, given the cook a few directions, and then gone out on the lawn at the back of the house to have a quiet read and enjoy the soft air and sunshine of that lovely May morning. She lay down in a hammock chair in the shade of a fine old cedar at the bottom of the lawn, and began to read, and soon she began to dream. The news in the papers, even the most responsible of them, had been very serious. The shadow of war was once more rising in the East—war which, if it came, England could scarcely escape, and if it did Someone would have to go and fight in that most perilous of all forms of battle, torpedo attack.

The book she had taken with her was one of exceedingly clever verse written years before by just such another as herself; a girl, beautiful, learned, and yet absolutely womanly, and endowed, moreover, with that gift so rare among learned women, the gift of humour. Long ago, this girl had taken the fever in Egypt, and died of it; but before she died she wrote a book of poems and verses, which, though long forgotten—if ever known—by the multitude, is still treasured and re-read by some, and of these Miss Nitocris was one. Just now the book was open at the hundred and forty-third page, on which there is a portion of a poem entitled *Natural Selection*.

Miss Nitocris' eyes alternately rested on the page for a few moments and then lifted and looked over the lawn towards the open French windows. The verses ran thus:

> "But there comes an idealless lad,
> With a strut, and a stare, and a smirk;
> And I watch, scientific though sad,
> The Law of Selection at work.

"Of Science he hasn't a trace,
He seeks not the How and the Why,
But he sings with an amateur's grace
And he dances much better than I.

"And we know the more dandified males
By dance and by song win their wives —
'Tis a law that with Aves *prevails,*
And even in Homo *survives."*

"Just my precious papa's ideas!" she murmured, with a toss of her head, and something like a little sniff. "What a nuisance it all is! Aristocracy of intellect, indeed! Just as if any of us, even my dear Dad, if he *is* considered one of the cleverest and most learned men in Europe, were anything more than what Newton called himself—a little child picking up pebbles and grains of sand on the shore of a boundless and fathomless ocean, and calling them knowledge. I'm not quite sure that that's correct, but it's something like it. Still, that's not the question. How on earth am I to tell poor Mark? Oh dear! he'll have to be 'Mr Merrill' now, I suppose. What a shame! I've half a mind to rebel, and vindicate the Law of Selection at any price. Ah, there he is. Well, I suppose I've got to get through it somehow."

As she spoke, one of the French windows under the verandah opened, and a man in a panama hat, Norfolk jacket and knickerbockers, came out and raised his hat as he stepped off the verandah.

With a sigh and a frown she closed the book sharply, got up and tossed it into the chair. No daintier or more desirable incarnation of the eternal feminine could have been imagined than she presented as she walked slowly across the lawn to meet the man whom the Law of Selection had designated as her natural mate, and whom her father, for reasons presently to be made plain, had forbidden her to marry on pain of exile from his affections for ever.

The face he turned towards her as she approached was not exactly handsome as an artist or some women would have defined the word, but it was strong, honest, and open—just the sort of face, in short, to match the broad shoulders, the long, cleanly-shaped, athletic limbs, and the five feet eleven of young, healthy manhood with which Nature had associated it.

A glance at his face and another one at him generally would, in spite of the costume, have convinced any one who knows the genus that Mark

Merrill was a naval officer. He had that quiet air of restrained strength, of the instinctive habit of command which somehow or other does not distinguish any other fighting man in the world in quite the same degree. His name and title were Lieutenant-Commander Mark Gwynne Merrill, of His Majesty's Destroyer *Blazer*, one of the coolest-headed and yet most judiciously reckless officers in the Service.

There was a light in his wide-set, blue-grey eyes, and a smile on his strong, well-cut lips which were absolutely boyish in their anticipation of sheer delight as she approached; and then, after one glance at her face, his own changed with a suddenness, which, to a disinterested observer, would have been almost comic.

"I'm awfully sorry, Mark," she began, in a tone which literally sent a shiver—a real physical shiver—through him, for he was very, very much in love with her.

"What on earth is the matter, Niti?" he said, looking at the fair face and downcast eyes which, for the first time since he had asked the eternal question and she had answered it according to his heart's desire, had refused to meet his. "Let's have it out at once. It's a lot better to be shot through the heart than starved to death, you know. I suppose it's something pretty bad, or you wouldn't be looking down at the grass like that," he continued.

"Oh, it's—it's—it's a *beastly* shame, that's what it is, so there!" And as she said this Miss Nitocris Marmion, B.Sc., stamped her foot on the turf and felt inclined to burst out crying, just as a milkmaid might have done.

"Which means," said Mark, pulling himself up, as a man about to face a mortal enemy would do, "that the Professor has said 'No.' In other words, he has decided that his learned and lovely daughter shall not, as I suppose he would put it, mate with an animal of a lower order—a mere fighting-man. Well, Miss Marmion——"

"Oh, don't; *please* don't!" she exclaimed, almost piteously, dropping into a big wicker armchair by the verandah and putting her hands over her eyes.

He had an awful fear that she was going to cry, and, as the Easterns say, he felt his heart turning to water within him. But her highly trained intellect came to her aid. She swallowed the sob, and looked up at him with clear, dry eyes.

"It isn't quite that, Mark," she continued. "You know I wouldn't stand anything like that even from the dear old Dad. Much as I love him, and even, as you know, in some senses almost worship him, it isn't that. It's this theory of heredity of his—this scientific faith—bigotry, I call it, for it is just the same to him as Catholicism was to the Spaniards in the sixteenth century. In fact, I told him the other night that he reminded me of the Spanish grandee whose daughters were convicted of heresy by the Inquisition, and who showed his devotion to the Church by lighting the faggots which burned them with his own hands."

"And what did he say to that?" said the sailor, not because he wanted to know, but because there was an awkward pause that needed filling.

"I would rather not tell you, Mark, if you don't mind," she said slowly and looking very straightly and steadily at him. "You know—well, I needn't tell you again what I've told you already. You know I care for you, and I always shall, but I cannot—I dare not—disobey my father. I owe all that I ever had to him. He has been father, mother, teacher, friend, companion—everything to me. We are absolutely alone in the world. If I could leave him for anybody, I'd leave him for you, but I won't disobey him and break his heart, as I believe I should, even for you."

"You're perfectly right, Niti, perfectly," said Commander Merrill, in a tone of steady conviction which inspired her with an almost irresistible impulse to get up and kiss him. "You couldn't honestly do anything else, and I know the shortest way to make you hate me would be to ask you to do that something else. But still," he went on, thrusting his hands into the pockets of his Norfolk jacket, "I do think I have a sort of right to have some sort of explanation, and with your permission I shall just ask him for one."

"For goodness' sake, don't do that, Mark—don't!" she pleaded. "You might as well go and ask a Jewish Rabbi why he wouldn't let his daughter marry a Christian. Wise and clever as he is in other things, poor Dad is simply a fanatic in this, and—well, if he did condescend to explain, I'm afraid you might mistake what he would think the correct scientific way of putting it, for an insult, and I couldn't bear to think of you quarrelling. You know you're the only two people in the world I—I—Oh dear, what *shall* I do!"

It was at this point that the Law of Natural Selection stepped in. Natural laws of any sort have very little respect for the refinements of what mortals are pleased to call their philosophy. Professor Marmion was a very great man—some men said he was the greatest scientist of his age—but at this moment he was but as a grain of sand among the wheels of the mighty machine which grinds out human and other destinies.

Commander Merrill took a couple of long, swift strides towards the chair in which Nitocris was leaning back with her hands pressed to her eyes. He picked her up bodily, as he might have picked a child of seven up, put her protesting hands aside, and slowly and deliberately kissed her three times squarely on the lips as if he meant it; and the third time her lips moved too. Then he whispered:

"Good-bye, dear, for the present, at any rate!"

After which he deposited her tenderly in the chair again, and, with just one last look, turned and walked with quick, angry strides across the lawn and round the semi-circular carriage-drive, saying some things to himself between his clenched teeth, and thinking many more.

A few yards outside the gate he came face to face with the Professor.

"Good-morning, sir," said Merrill, with a motion of his hand towards his hat.

"Oh, good-morning, Mr Merrill," replied the Professor a little stiffly, for relations between them had been strained for some considerable time now. "I presume you have been to the house. I am sorry that you did not find me at home, but if it is anything urgent and you have half an hour to spare——"

He stopped in his speech, silenced by a shock of something like shame. He was prevaricating. He knew perfectly well that "it" was the most urgent errand a man could have, next to his duty to his country, that had brought the young sailor to his house. Twenty-four hours ago he would not have noticed such a trifle: but it was no trifle now; for to his clearer vision it was a sin, an evasion of the immutable laws of Truth, utterly unworthy of the companion of Nitocris the Queen in that other existence which he had just left.

"You have seen Niti, I suppose?" he continued, with singular directness.

"Yes," replied Merrill. "You will remember that the week was up this morning, and so I called to learn my fate, and your daughter has told me. I presume that your decision is final, and that, therefore, there is nothing more to be said on the subject."

"My decisions are usually final, Mr Merrill, because I do not arrive at them without due consideration. I am deeply grieved, as I have told you before, but my decision is a deduction from what I consider to be an unbreakable chain of argument which I need not trouble you with. Personally and socially, of course, it would be impossible for me to have the slightest objection to you. In fact, apart from your execrable fighting profession, I like you; but otherwise, as you know, I cannot help looking at you as the survival of an age of barbarism, a hark-back of humanity, for all the honour in which that trade is held by an ignorant and deluded world; and so for the last time it is my painful task to tell you that there can be no union between your blood and mine. Outside that, of course, there is no reason why we should not remain friends."

"Very well, sir," replied Merrill, "I have heard your decision, and Miss Marmion has told me she is resolved to abide by it; I should be something less than a man if I attempted to alter her resolve. We are ordered on foreign service this week, and so for the present, good-bye."

He lifted his hat, turned away and walked down the road with teeth clenched and eyes fixed straight in front of him, and a shade of grey under the tan of his skin.

The Professor looked after him for a few moments and turned in at the gate, saying:

"It's a great pity in some ways—many ways, in fact. He's a fine young fellow and a thorough gentleman, and I'm afraid they're very fond of each other, but of course to let Niti marry him would be the negation of the belief and teaching of more than half a lifetime. I hope the poor girl won't take it too keenly to heart. I'm afraid he seems rather hard hit, poor chap, but of course there's no help for it. Just fancy me the father-in-law of a fighting man, and the grandfather of what might be a brood of fighters! No, no; that is quite out of the question."

CHAPTER VII
MOSTLY POSSIBILITIES

The Professor went into the garden feeling just a trifle uncomfortable. He not only loved his daughter dearly, but he also had a very deep and well-justified respect for her intellect and scholarly attainments. Her unfortunate love for a man whom he honestly believed to be a totally unfit mate for her was the only shadow that had ever drifted between them since she had become, not only his daughter, but his friend and companion, and the enthusiastic sharer of his intellectual pursuits. Of course, anything like a scene was utterly out of the question; but there is a silence more eloquent than words, and it was that that he was mostly afraid of.

He found her walking up and down the lawn with her hands behind her back. She was a little paler than usual, and there was a shadow in her eyes. She came towards him, and said quite quietly:

"Mr Merrill has been here, Dad, to say good-bye. I told him, and so we have said it."

The simple words were spoken with a quiet and yet tender dignity which made him feel prouder than ever of his daughter and all the more sorry for her.

"I met him just outside the gate, Niti," he replied, looking at her through a little mist in his eyes, "He spoke most honourably, and like the gentleman that he is. I hope you will believe me——"

"I believe you in everything, Dad," she said quickly; "and since the matter is ended, it will only hurt us both to say any more about it. Now, I have some news," she continued, in a tone whose alteration was well assumed.

"Ah! and what is that, Niti?" he asked, looking up at her with a smile of relief.

"It's something that I hope you will be able to get some of your solemn fun out of. One of the items in the 'Social Intelligence' to-day states that your old friend, Professor Hoskins van Huysman, and his wife and daughter have come to London, and will stay ten days before 'proceeding' to Paris and the South of France, and so, of course, they will be here for your lecture, and naturally he will not resist the temptation of making one of your audience."

"Van Huysman!" exclaimed the Professor. "That Yankee charlatan, confound him! I shouldn't wonder if he had the impudence to take part in the discussion afterwards."

"Then," laughed Nitocris, "you must take care to have all your heavy guns ready for action. But, of course, Dad, you won't let your—well, your scientific feelings get mixed up with social matters, will you? Because, you know, I like Brenda very much; she's the prettiest and brightest girl I know. You know, she can do almost anything, and yet she's as unaffected——"

"As some one else we know," interrupted the Professor with another smile.

"And then, you know, Mrs van Huysman," continued Nitocris with a little flush, "is such a dear, innocent, good-natured thing, so good-hearted and so deliciously American. Of course, you can fight with the Professor as much as you like in print, and in lecture halls—I know you both love it—but you'll still be friends socially, won't you?"

"Which, of course, means garden-parties and river trips, and similar frivolities that learned young ladies love so much. You needn't trouble about that, Niti. I shall not allow my zeal for scientific truth to interfere with your social pleasures, you may be quite sure. Science, as you know, has nothing to do with what we call Society, except as one of the most curious phenomena of Sociology. Drive into town whenever you like and see them. Present my respectful compliments, and ask them to dinner, or whatever you like. And now I must get to my work—I've only three more days, and my notes are not anything like complete."

"Very well, Dad; I think I'll telephone them—they're stopping at the Savoy—extravagant people!—to say that I'll run in this afternoon and have tea. Oh! and, by the way," she added, as he turned towards the house, "there's another item. Lord Leighton has been called home suddenly on some business, and will be here the day after to-morrow."

"Oh! indeed," said the Professor, pausing. "Well, I shall be delighted to see him—but I don't know what I shall have to say to him about that Mummy."

Nitocris turned away towards her chair with a faint smile on her lips. With a woman's rapid intuition, she had seen a glimmer of hope in the conjunction of these two announcements. Although Professor van Huysman's personal fortune was not as great as his attainments or his fame, Brenda would be very rich, for her mother was the only sister of a widower whose sole interest and occupation in life was piling up dollars. He had dollars in everything, from pork and lumber to canned goods, and her own father's scientific inventions, and Brenda was the bright particular star of his affections.

On the other hand, Lord Leighton, son and heir of the invalid Earl of Kyneston, was a fairly well-to-do young nobleman, good-looking, a scholar, and a good sportsman, who had done brilliantly at Cambridge, and then devoted himself to Egyptian exploration with a whole-souled ardour which had quickly won Professor Marmion's heart, and a ready consent to his "trying his luck" with his daughter to boot. This had not a little to do with the present unfortunate condition of her own love affairs.

She had already refused Lord Leighton, letting him down, of course, as gently as possible, but withal firmly and uncompromisingly. Who could better console him than this beautiful and brilliant American girl, and what would better suit that lovely head of hers than an English coronet which was bright with the untarnished traditions of five hundred years?

Wherefore, then and there, Miss Nitocris Marmion, Bachelor of Science, Licentiate of Literature and Art, and Gold-Medallist in Higher Mathematics at the University of London, decided upon her first experiment in match-making.

When the Professor got into his study and shut the door, there was a curious smiling expression upon his refined, intellectual features. Instead of sitting down to his desk, he lit a pipe and began walking up and down the room, communing with his own soul in isolated sentences, as was his wont when he was trying to arrive at any difficult decision.

In order to appreciate his deliberations and their result, it will be necessary to say that Professor Hoskins van Huysman was one of the most distinguished physicists in America, and he had also gained distinction in applied mathematics. In addition to this, he was the inventor of many marvellous contrivances for the demonstration and measurement of the more obscure physical forces. His official position was that of Lecturer and Demonstrator in Physical Science in Harvard University.

He and Professor Marmion had been deadly opponents in the field of controversy for years. The latter had once detected an error in a very learned monograph which he had published in the *Scientific American* on the "Co-

Relation of the Etheric Forces in the Phenomena of Light and Heat," and of course he had never forgiven him. From that day forth a relentless duel of wits between them had continued. Every essay, monograph, or book that the one published, the other criticised with cold but ruthless severity, to the great delectation of the scientific world, if not to the clarification of its atmosphere.

Socially, they were cordial acquaintances, if not friends. What they really thought of each other was known only to themselves and to their immediate domestic circles.

Naturally Professor Marmion was well aware that his elevation to the higher plane of N^4 gave him an enormous advantage over his adversary, for now he could, if he chose, smite him hip and thigh, in a strictly scientific sense, and reduce him to utter confusion and public ridicule, and the question which he had come to discuss with himself was: In how far, if at all, was he justified in so using the extra-human powers with which he had been endowed?

The moment that he began to do this he became conscious of another curious complication of his recent development. On the higher plane he had argued the matter out with no more emotion than a calculating machine would have betrayed, and he had come to a conclusion that was absolutely luminous and just: but now that he came to argue the same question on the lower plane he found that he was doing it under human limitations, and therefore with human feelings.

"No," he said in the peculiar low, musing tone which was habitual to him during these monologues, "no; after all, I do not see that there would be any harm in that. Wrong, nay, sinful it would undoubtedly be to prove to demonstration that religious, social, and physical laws, may, under certain changing circumstances, be both true and false at the same time. I am, or was—or whatever it is—perfectly right in considering that to deliberately produce such a chaos as that would do would be the most colossal crime that a man could commit against humanity, as far as this plane is concerned, but there can be no harm in making a few mathematical experiments."

He took a few more turns up and down the room, pulling slowly at his pipe, and with his mind not wholly unoccupied with speculations as to what Professor Van Huysman's feelings might be if he were watching the said experiments. Then he began again:

"At the worst I shall only be carrying certain investigations a few steps farther, and developing theories which have been seriously discussed by the hardest-headed scholars in the world. Both the Greek and the Alexandrian philosophers speculated on the possibility of a state of four dimensions; and didn't Cayley, before this very Society, deliberately say that at the present rate of progress in the Higher Mathematics, the eye of Intellect might ere long see across the border of tri-dimensional space?

"Surely I cannot do any very great harm by carrying his arguments to their logical conclusions—if I can. Of course, physical demonstrations would never do: I should frighten my brilliant and learned audience out of its seven senses; but, as for mere mathematics—well, I may make them stare, and set a good many highly-respected brains—my gifted friend Huysman's, among them—working pretty hard. Of course, he will be especially furious, but there's no harm in that either. Yes, I shall certainly do it. If he can't understand my demonstrations, that's not my concern."

He went and sat down at his desk, still smiling, and went very carefully through the notes he had already made, and then through Professor Hartley's letter, and his speculations on the Forty-Seventh Proposition. This done, he plunged into a fresh vortex of figures, and symbols, and diagrams, in which he remained for the next two hours, his mind hovering, as it were, over the borderland which at once divides and unites the higher and the lower planes. When he returned to earth, the dreamy, abstracted look faded away from his face; his eyes lit up, and the pleasant smile came back.

He opened the middle drawer in his desk, and took out the first page of the fair copy of his notes, which Nitocris had made for him—thinking the while how easy it would have been for him in the state of N^4 to take it out without opening the drawer at all—and looked at it. It was headed:

"RECENT PROGRESS IN THE HIGHER MATHEMATICS."

He crossed the title out carefully, and wrote above it:

"AN EXAMINATION OF SOME SUPPOSED MATHEMATICAL IMPOSSIBILITIES."

"There," he murmured, as he put the sheet back; "I think that such a theme, adequately treated, will considerably astonish my learned friends in general, and my esteemed critic, Van Huysman, in particular."

From which remark it will be gathered that Franklin Marmion had certainly recrossed the dividing line between the two Planes of Existence.

CHAPTER VIII
MISS BRENDA ARRIVES, AND PHADRIG THE EGYPTIAN PROPHESIES

"Now, this is just too sweet of you, Niti, to come so soon after we got here. In five minutes more I should have written you a note, asking you and the Professor to come and take lunch with us to-morrow, and here you've anticipated me, so we have the pleasure of seeing you all the sooner."

These were the words with which Miss Brenda van Huysman greeted Nitocris as she entered the drawing-room of the suite of apartments which formed her home for the time being in London. I say her home advisedly, because, although her father and mother also occupied it, she was virtually, if not nominally, mistress undisputed of the splendid camping-place.

She was an almost perfect type of the highly developed, highly educated American girl of to-day, a marvellous compound of intense energy and languorous grace. She had done as brilliantly at Vassar as Nitocris had done at Girton and London, and she had also rowed stroke in the Ladies' Eight, and was champion fencer of the College. Yet as far as her physical presence was concerned, she was just a "Gibson Girl" of the daintiest type—fair-skinned, blue-eyed, golden-haired—her hair had a darker gleam of bronze in it in certain lights—exquisitely moulded features which seemed capable of every sort of expression within a few changing moments, and a poise of head and carriage of body which only perfect health and the most scientific physical training can produce. In a word, she was one of those miraculous developments of femininity which Nature seems to have made a speciality for the particular benefit of the younger branch of the Anglo-Saxon race. As for her dress—well, the shortest and best way to describe that is to say that it exactly suited her.

As she spoke, and their hands met, Mrs van Huysman got up and came towards them, saying:

"Good afternoon, Miss Marmion. We were real glad to get your 'phone, and it's good to see you again. How's the Professor? Too busy to come with you, I suppose, as usual. We see he's going to lecture before the Royal Society on the tenth, and I reckon we shall all be there to listen to him. I shouldn't wonder but there'll be trouble as usual between him and my husband. It seems a pity that two such clever men should waste so much time in scrapping over these scientific things, which don't seem to matter half a cent, anyhow."

"Oh, I don't know," laughed Nitocris, as they shook hands. "You see, Mrs van Huysman, *they* do think it matters a great deal, and, besides, I'm quite sure that they both enjoy it very thoroughly. It's their way of taking recreation, you see, just as a couple of pitmen will try and pound one another to pieces, just for the fun of the thing. It's only a case of intellectual fisticuffs, after all."

"Why, certainly," said Brenda, as she rang for tea; "I'm just sure that Poppa never has such a good time as when he thinks he's tearing one of Professor Marmion's theories into little pieces and dancing on them, and I shouldn't wonder if Professor Marmion didn't feel about the same."

"I dare say he does," said Nitocris, remembering what had happened in the morning; "it's only one of the thousand unexplained puzzles of human nature. As you know, my father hates fighting in the physical sense with a hatred which is almost fanatical, and yet, when it comes to a battle of wits, he's like a schoolboy in a football match."

"It's just another development of the same thing," said Brenda. "Man was born a fighting animal, and I guess he'll remain one till the end of time; and with all our progress in civilisation and science, and all that, the man who doesn't enjoy a fight of some sort isn't of very much account. Now, here's tea, which is just now a more interesting subject. Sit down, and we'll talk about vanities. I'm just perishing to see what Regent Street and Bond Street are like. I don't think I've spent ten dollars in London yet. I'm twenty-two to-morrow, Niti, and my grandfather, who is just about the best grandfather a girl ever had, cabled across to the Napier people, and they've sent round the dandiest six-cylinder, thirty-horse landaulette that you ever saw, even in Central Park, and a driver to match—only I shan't have much use for him, except to look after the automobile. I'll run you round in her after tea, and you can reintroduce me to the stores—I mean shops; I forgot we were in London."

Mrs van Huysman, as usual, took a back seat while her daughter dispensed tea, and did most of the talking. She was a lady of moderate proportions, and, unlike a good many American women, she had kept her good looks until very close on fifty. She was full of shrewd common sense, but she had been born in a different generation and in a different grade of life, and therefore her attire inclined rather to magnificence than to elegance, in spite of her daughter's restraining hand and frankly expressed counsel. She had a profound respect for her husband's attainments without in the least understanding them, and she very naturally held an unshakable belief that no quite ordinary woman, as she called herself, had ever been miraculously blessed with such a daughter as she had.

Nitocris was just beginning her second cup of tea when the door opened and her father's foeman in the arena of Science came in. He was the very antithesis of Professor Marmion; a trifle below middle height, square-shouldered and strongly built, with thick, iron-grey hair, and somewhat heavy features which would have been almost commonplace but for the broad, square forehead above them, and the brilliant steel-grey eyes which glittered restlessly under the thick brows, and also a certain sensitiveness about the nostrils and lips which seemed curiously out of keeping with the strength of the lower jaw. His whole being suggested a combination of restless energy and inflexible determination. If he had not been one of America's greatest scientists, he would probably have been one of her most ruthless and despotic Dollar Lords.

"Ah, Miss Marmion, good afternoon! Pleased to see you," he said heartily, as Nitocris got up and held out her hand. "Very kind of you to look us up so soon. How's the Professor? Well, I hope. I see he's scheduled for a lecture before the Royal Society. He's got something startling to tell us about, I hope. It's some time since we had anything of a scientific scrap between us."

"And therefore," said Nitocris, as she took his hand, "I suppose you are just dying for another one."

"Well, not quite dying," laughed the Professor. "Don't look half dead, do I? Just curious, that's all. You can't give me any idea of the subject, I suppose?"

"I could, Professor," she replied, with a malicious twinkle in her eye, because she had already had a talk with her father on the altered title of the lecture, "but if I did, you know, I should only, as we say in England, be spoiling sport. However, I don't think I shall be playing traitor if I tell you to prepare for a little surprise."

Professor van Huysman's manner changed instantly, and the warrior soul of the scientist was in arms.

"Oh yes! A surprise, eh?" he said, with something between a snort and a snarl in his voice. "Then I guess— —"

"Poppa, sit down and have some tea," said his daughter, quietly but firmly.

He sat down without a word, took his cup of tea and a slice of bread and butter; listened in silence as long as he could bear the entirely feminine conversation on a subject in which he hadn't the remotest interest, and then he put his cup down with a little jerk, got up with a bigger one, and said, holding out his hand to Miss Nitocris:

"Well, Miss Marmion, I shall have to say good afternoon. You see we've only just reached this side, and I've got quite a lot of things to attend to. Bring your father along to dinner to-morrow night, if you can; I shall be glad to meet him again. You needn't be afraid: we shan't shoot."

When he had gone, Brenda rang and ordered the motor-car to be ready in half an hour. Then they finished their tea and talk, and Brenda and Nitocris went and put on their wraps—not the imitation of the mediæval armour which is used for serious motor-driving, but just dust-cloaks and mushrooms, both of which Brenda lent to her friend. As they came back through the drawing-room, she said to her mother:

"Well, Mamma, the car's ready, I believe. Won't you join us in a little run round town?"

"When I want to take a run into the Other World in one of those infernal machines of yours, Brenda," said her mother, with a mild touch of sarcasm in her tone, "I'll ask you to let me come. This afternoon I feel just a little bit too comfortable for a journey like that."

"It's a curious thing," said Brenda, as they were going down in the lift, "Mamma's as healthy a woman as ever lived, and she's American too, and yet I believe she'd as soon get on top of a broncho as into an automobile."

The car was waiting for them in the courtyard under the glass awning. A smart-looking young *chauffeur* in orthodox costume touched his cap and set the engine going. The gold-laced porters handed them into the two front seats, and the *chauffeur* effaced himself in the *tonneau*. Miss Brenda put one hand on the steering-wheel and the other on the first speed lever, and the car slid away, as though it had been running on ice, towards the great arched entrance.

As they turned to the left on their way westward, a shabbily dressed man and woman stepped back from the roadway on to the pavement. For a moment they stared at the car in mute astonishment; then the man gripped the woman tightly by the arm and led her away out of the ever-passing throng, whispering to her in Coptic:

"Did'st thou see her, Neb-Anat—the Queen—the Queen in the living flesh sitting there in the self-mover, the devil-machine? To what unholy things has she come—she, the daughter of the great Rameses! But it may be that she is held in bondage under the spell of the evil powers that created these devil-chariots which pant like souls in agony and breathe with the breath of Hell. She must be rescued, Neb-Anat."

"Rescued?" echoed the woman, in a tone that was half scorn and half fear. "Is it so long ago that thou hast forgotten how we tried to rescue her mummy from the hands of these infidels? Now, behold, she is alive again, living in the midst of this vast, foul city of the infidels, clothed after the fashion of their women, and yet still beautiful and smiling. Pent-Ah, didst thou not even see her laugh as she rode past us? Alas! I tell thee that our Queen is laid under some awful spell, doubtless because she has in some way incurred the displeasure of the High Gods, and if that is so, not even the Master himself could rescue her. What, then, shall we do?"

"Thy saying is near akin to blasphemy, Neb-Anat," he murmured in reply, "and yet there may be a deep meaning in it. Nevertheless, to-night, nay, this hour, the Master must know of what we have seen."

They walked along, conversing in murmurs, as far as Waterloo Bridge, then they turned and crossed it and walked down Waterloo Road into the Borough Road, and then turned off into a narrow, grimy street which ended in a small court whose three sides were formed of wretched houses, upon which many years of misery, poverty, and crime had set their unmistakable stamp. They crossed the court diagonally and entered a house in the right-hand corner. They went up the worn, carpetless stairs with a rickety handrail on one side and the torn, peeling paper on the other, and stopped before a door which opened on to a narrow landing on the first floor. Pent-Ah knocked with his knuckles on the panel, first three times quickly, and then twice slowly. Then came the sound of the drawing of a bolt, and the door opened.

They went in with shuffling feet and crouching forms, and the woman closed the door behind her. A tall, gaunt, yellow-skinned man, his head perfectly bald and the lower part of his face covered with a heavy white beard and moustache, faced them. His clothing was half Western, half Oriental. A pair of thin, creased, grey tweed trousers met, or almost met, a pair of Turkish slippers, showing an inch of bare, lean ankle in between. His body was covered with a dirty yellow robe of fine woollen stuff, whose ragged fringe reached to his knees, and a faded red scarf was folded twice round his neck, one end hanging down his breast and the other down his back. As Pent-Ah closed the door and bolted it, he said to him in Coptic:

"So ye have returned! What news of the Queen? For without that surely ye would not have dared to come before me."

He spoke the words as a Pharaoh might have spoken them to a slave, and as though the bare, low-ceiled, shabby room, with its tawdry Oriental curtains and ornaments, had been an audience-chamber in the palace of Pepi in old Memphis, for this was he who had once been Anemen-Ha, High Priest of Ptah, in the days when Nitocris was Queen of the Two Kingdoms.

"We have seen her once more, Lord," said Pent-Ah, "scarce an hour ago, dressed after the fashion of these heathen English, and seated in a devil-chariot beside another woman, as fair almost as she. It is true, Lord, even as we said, that our Lady the Queen is in the flesh again, and yet she knows us not. It may be that the High Gods have laid some spell upon her."

"Spell or no spell, the mission which is ours is the same," was the reply. "It is plain that a miracle has been worked. The Mummy which we—I as well as you—were charged to recover and restore to its resting-place, has vanished. The Queen has returned to live yet another life in the flesh, but the command remains the same. Mummy or woman, she shall be taken back to her ancient home to await the day when the Divine Assessors shall determine the penalty of her guilt. The task will be hard, yet nothing is impossible to those who serve the High Gods faithfully. Ye have done well to bring me this news promptly. Here is money to pay for your living and your work. Watch well and closely. Know every movement that the Queen makes, and every day inform me by word or in writing of all her actions. On the fourth day from now come here an hour before midnight. Now go."

He counted out five sovereigns to Pent-Ah. Their glitter contrasted strangely with the shabby squalor of the room and the poverty of his own dress, but he gave them as though they had been coppers. Pent-Ah took them with a low obeisance, and dropped them one by one into a pocket in a canvas belt which he wore under his ragged waistcoat. Neb-Anat looked at them greedily as they disappeared.

"The Master's commands shall be obeyed, and the High Gods shall be faithfully served," said Pent-Ah, as he straightened himself up again. "From door to door the Queen shall be watched, and, if it be permitted, Neb-Anat shall become her slave, and so the watch shall be made closer. Is not that so, Neb-Anat?"

"The will of the Master is the law of his slave," she replied, sinking almost to her knees.

"It is enough," replied the Master, who was known to the few who knew him as Phadrig Amena, a Coptic dealer in ancient Egyptian relics and curios in a humble way of business. "Serve faithfully, both of you, and your reward shall not be wanting. Farewell, and the peace of the High Gods be on you."

When they had gone he sat down to the old bureau, took out a sheaf of papers, some white and new, others yellow-grey with age, and yet others which were sheets of the ancient papyrus. The writing on these was in the old Hermetic character; of the rest some were in cursive Greek and some in Coptic. A few only were in English, and about half a dozen in Russian. He read them all with equal ease, and although he knew their contents almost by heart, he pored over them for a good half-hour with scarcely so much as a movement of his lips. Then he put them away and locked the drawer with one of a small bunch of curiously shaped keys which were fastened round his waist by a chain. When he had concealed them in his girdle, he got up and began to pace the floor of the miserable room with long, stately, silent steps as though the dirty, cracked, uneven boards had been the gleaming squares of alternate black and white marble of the floor of the Sanctuary in the now ruined Temple of Ptah in old Memphis. Then, after a while, with head thrown proudly back and hands clasped behind him, he began to speak in the Ancient Tongue, as though he were addressing some invisible presence.

"Yes, truly the Powers of Evil and Darkness have conquered through many generations of men, but the days of the High Gods are unending, and the climax of Fate is not yet. Not yet, O Nitocris, is the murderous crime of thy death-bridal forgotten. The souls of those who died by thy hand in the banqueting chamber of Pepi still call for vengeance out of the glooms of Amenti. The thirst of hate and the hunger of love are still unslaked and unsatisfied. I, Phadrig, the poor trader, who was once Anemen-Ha, hate thee still, and the Russian warrior-prince, who was once Menkau-Ra, shall love thee yet again with a love as fierce as that of old, and so, if the High Gods permit, between love and hate shalt thou pass to the doom that thou hast earned."

He paused in his walk and stood staring blankly out of the grimy little window with eyes which seemed to see through and beyond the smoke-blackened walls of the wretched houses opposite, and away through the mists of Time to where a vast city of temples and palaces lay under a cloudless sky beside a mighty slow-flowing river, and his lips began to move again as those of a man speaking in a dream:

"O Memphis, gem of the Ancient Land and home of a hundred kings, how is thy grandeur humbled and thy glory departed! Thy streets and broad places which once rang with the tramp of mighty hosts and echoed with the songs of jubilant multitudes welcoming them home from victory are buried under the drifting desert sands; in the ruins of thy holy temples the statues of the gods lie prone in the dust, and the owl rears her brood on thy crumbling altars, and hoots to the moon where once rose the solemn chant of priests and the sweet hymns of the Sacred Virgins; the jackal barks where once the mightiest monarchs of earth gave judgment and received tribute; thy tombs are desecrated, and the mummies of kings and queens and holy men have been ravished from them to adorn the unconsecrated halls of the museums of ignorant infidels; the heel of the heathen oppressor has stamped the fair flower of thy beauty into the deep dust of defilement. Alas, what great evil have the sons and daughters of Khem wrought that the High Gods should have visited them with so sore a judgment! How long shall thy bright wings lie folded and idle, O Necheb, Bringer of Victory?"

A deep sigh came from his heaving breast as he turned away and began his walk again. Soon he spoke again, but now in a changed voice from which the note of exaltation had passed away:

"But it is of little use to brood over the lost glories of the past. Our concern is with that which is and that which may—nay, shall be. Who is this Franklin Marmion, this wise man of the infidels? Who is he, and who was he—since, by the changeless law of life and death, each man and woman is a deathless soul which passes into the shadows only to return re-garbed in the flesh to live and work through the interlocked cycles of Eternal Destiny? Was he—ah Gods! was *he* once Ma-Rimōn, whose footsteps in the days that are dead approached so nearly to the threshold of the Perfect Knowledge, while mine, doubtless for the sin of my longing for mere earthly power and greatness, were caught and held back in a web of my own weaving? And, if so, has he attained while I have lost?

"What if that strange tale which Pent-Ah and Neb-Anat told me of their visit to his house—told, as I thought, to hide their failure under a veil of lies—was true? If so, then he has passed the threshold and taken a place only a little lower than the seats of the gods, a place that I may not approach, barred by the penalty of my accursed folly and pride! Ah well, be it so or be it not, are not the fates of all men in the hands of the High Gods who see all things? We see but a little, and that little, with their help, we must do according to the faith and the hope that is in us."

At this moment there came a knock at the door. It opened at his bidding, and a dirty-faced, ragged-frocked little girl shuffled into the room holding out a letter in her hard, grimy, claw-like hand.

"'Ere's somethin' as has just come for you, Mister Phadrig. Muvver told me ter bring it up, and wot'll yer want for supper, and will yer give me the money?" she said in a piping monotone, still holding out her hand after he had taken the letter. He gave her sixpence, saying:

"Two eggs and some bread. I will make my coffee myself."

She took the coin and shuffled out quickly, for she went not a little in awe of this dark-faced foreign man from mysterious regions beyond her ken, who was doubtless a magician of some sort, and could kill her or change her into a rat by just breathing on her, if he wanted to.

Meantime Nitocris and Brenda were having what the latter called "a perfectly lovely time" in Regent Street and Bond Street and other purlieus of that London paradise which the genius of commerce has created for the delight of his richest and most lavish-handed votaries. Brenda spent her ten dollars and a few thousands more, and then, as it was getting on to dinnertime and Nitocris absolutely refused to let her father eat his meal alone, she ran her out to Wimbledon at a speed for which a mere man would have inevitably been fined, asked herself to dinner, and made herself entirely delightful to the Professor.

But in spite of all her cunning wiles and winning ways she left in absolute ignorance of the subject of the forthcoming lecture.

CHAPTER IX
"THE WILDERNESS,"WIMBLEDON COMMON

The little estate on Wimbledon Common, which had been in Professor Marmion's family for three generations, was called "The Wilderness." The house was of distinctly composite structure. Tradition said that it had been a royal hunting lodge in the days when Barnes and Putney and Wimbledon were tiny hamlets and the Thames flowed silver-clear through a vast, wild region of forest and gorse and heather, and the ancestors of the deer in Richmond Park browsed in the shade of ancient oaks and elms and beeches, and antler-crowned monarchs sent their hoarse challenges bellowing across the open spaces which separated their jealously guarded domains.

Generation by generation it had grown with the wealth and importance of its owners, as befits a house that is really a home and not merely a place to live in, until it had become a quaint medley of various styles of architecture from the Elizabethan to the later Georgian. Thus it had come to possess a charm that was all its own, a charm that can never belong to a house that has only been built, and has not grown. Its interior was an embodiment in stone and oak and plaster of cosy comfort and dignified repose, and, though it contained every "modern improvement," all was in such perfect taste and harmony that even the electric light might have been installed in the days of the first James.

The Professor inhabited the northern wing, reputed to have been the original lodge in which kings and queens and great soldiers and statesmen had held revel after the chase, and tradition had endowed it with a quite authentic ghost: which was that of a fair maiden who had been decoyed thither to become the victim of royal passion, and who, strangely enough, poisoned herself in her despair, instead of getting herself made a duchess and founding the honours of a noble family on her own dishonour.

Although, as I have said, quite authentic, for the Professor had seen her so often that he had come to regard her with respectful friendship, the Lady Alicia was not quite an orthodox ghost. She did not come at midnight and wail in distressing fashion over the scene of her sad and shameful death. She seemed to come when and where she listed, whether in the glimpses of the moon or the full sunlight of mid-day. She never passed beyond the limits of the old lodge, and never broke the silence of her coming and goings. None of the present inhabitants of "The Wilderness" had seen her save the Professor, but Nitocris had often shivered with a sudden chill when she chanced to be in her invisible presence, and at such times she would often say to her father:

"There is something cold in the room, Dad. I suppose your friend the Lady Alicia is paying you a visit. I do wish she would allow me to make her acquaintance."

And to this he would sometimes reply with perfect gravity:

"Yes, she has just come in: she is standing by the window yonder." And this had happened so often that Nitocris, like her father, had come to regard the wraith, or astral body, as the Professor deemed it, of the unhappy lady almost as a member of the family. Of course, after he had passed the border into the realm of N^4, Franklin Marmion speedily came to look upon her visits as the merest commonplaces.

But as the unhappy Lady Alicia will have no part to play in the action of this narrative, her little story must be accepted as a perhaps excusable digression.

There were about four acres of comfortably wooded land about the house, of which nearly an acre had formed the pleasaunce of the old lodge. This was now a beautifully-kept modern garden, with a broad, gently-sloping lawn, whose turf had been growing more and more velvety year by year for over three centuries, and divided from it by a low box-hedge was another, levelled up and devoted to tennis and new-style croquet. The Old Lawn, as it was called, sloped away from a broad verandah which ran the whole length of the central wing and formed the approach to the big drawing-room and dining-room, and a cosy breakfast-room of early Georgian style, and these, with her study and "snuggery" and bedroom on the next floor, formed the peculiar domain of Miss Nitocris.

She and the Professor were just sitting down to an early breakfast on the morning of the garden-party, which had been arranged for the day but one after the arrival of the Huysmans, when the post came in. There were a good many letters for both, for each had many interests in life. The Professor only ran his eye over the envelopes and then put the bundle aside for consideration in the solitude of his own den. Nitocris did the same, picked one out and left the others for similar treatment after she had interviewed the cook about lunch and refreshments for the afternoon, and the butler on the subject of cooling drinks, for it promised to be a perfect English day in June—which is, of course, the most delicious day that you may find under any skies between the Poles.

She opened the one she had selected and skimmed its contents. Then her eyelids lifted, and she said:

"Oh!"

"What is the matter, Niti?" asked her father, looking up from his cutlet. "Nothing gone wrong with your arrangements, I hope."

"Oh dear, no," she replied, with something like exultation in her voice, "quite the reverse, Dad. This is from Brenda, and Brenda is an angel disguised in petticoats and picture hats. Listen."

.Then she began to read:

> "My dearest Niti,—I am going to take what I'm afraid English people would think a great liberty. The trouble is this: When the Professor (mine, I mean) was making his tour of the Russian Universities two years ago, he received a great deal of courtesy and help from no less a person than the celebrated Prince Oscar Oscarovitch—the modern Skobeleff, you know—who was very interested in Poppa's work, and took a lot of trouble to smooth things out for him. Well, the Prince, as of course you know, is in London now. He called yesterday, and when I mentioned your party, he said he was very sorry he had not the honour of your father's acquaintance as well as mine. The grammar's a bit wrong there, but you know what I mean. That, of course, meant that he wants to come; and, to be candid, I should like to bring him, for even an American girl here doesn't always get a Prince, and a famous man as well, to take around, so, as the time is so short, may we include him in our party? If you have forgiven me and are going to say 'yes,' I must tell you that the Prince would like to compensate for his intrusion— that's the way he puts it—by helping entertain your guests. It seems that he has met with a man who can work miracles, an Egyptian——"

At this point Professor Marmion looked up again suddenly with an almost imperceptible start, and, for the first time, took an interest in Miss Huysman's letter.

"— —named Phadrig. The Prince assures me that he is not a conjurer in the professional sense, and would be deeply insulted to be called one; also that no amount of money would induce him to give a display of his powers just *for* money. He will come to-day, if you like, and do wonderful things, which, from what the Prince says, will astonish and perhaps frighten us a bit, but only because the Prince once saved his life and got him out of a very bad place he had got into with a Turkish Pascha. Now, that is my little story. Please 'phone me as soon as you can so that I can let the Prince know. It will be just too sweet of you and the Professor to say 'yes.'

—Your devoted chum,

Brenda."

"Well, Dad," she asked, as she put the letter down, "what do you say?"

"Just what you want to say, my dear Niti," he replied, carefully spreading some marmalade on a triangle of toast "Personally, I must confess that I should rather like to see some of this so-called magician's alleged magic. I know that some of these fellows are extraordinarily clever, and I have no doubt that he will show us something interesting, if you care to see it."

"Then that settles it," said Nitocris, rising; "I will go and ring up the Savoy at once. Perhaps the Egyptian gentleman might be able to help you with that Forty-Seventh Proposition problem of Professor Hartley's."

"Perhaps," answered Franklin Marmion drily, and went on with his breakfast.

CHAPTER X
THE STAGE FILLS

The party which gradually assembled on the lawn about four was somewhat small, but very select. Nitocris had too much common sense and too much real consideration for her friends and acquaintances to get together a mere mob of well-dressed people of probably incompatible tastes and temperament, and call it a party. She disliked an elbowing crowd and a clatter of fashionably shrill tongues with all the aversion of a delicately developed sensibility. No consideration of rank or social power or wealth had the slightest weight with her when she was distributing cards of invitation, wherefore the said cards were all the more eagerly awaited by those who did, and did not, get them. The result of this in the present case was that, although every one accepted and came, rather less than fifty people had the run of the broad lawns and the leafy wilderness about them on that momentous afternoon.

The first of the arrivals was Professor Hartley, reputed to be the greatest mathematician in England. He was a large man with rather heavy features, lit up by alert grey eyes, a big, dome-like cranium, and a manner that was modest almost to diffidence. He brought his wife, a slim and somewhat stern-featured lady, who, in the domestic sense, kept him in his place with inflexible decision, and worshipped him in his professional capacity, and two pretty, well-dressed, and obviously well-bred daughters. Their carriage drew up, turned into the drive precisely at four. Punctuality was the Professor's one and only social vice.

Next came Commander Merrill in a hansom. This would be one of the very few meetings that he could hope for with his lost beloved—as he now sadly thought of her—before he put H.M.S. *Blazer* into commission, and so punctuality on his part was both natural and excusable. Then came a few more carriages containing very nice people with whom we have here but little concern; and then Miss Brenda, deeply regretting her beautiful Napier, with her father and mother in a very smart Savoy turn-out followed by a coronetted brougham drawn by a splendid pair of black Orloffs. This was followed by an equally smart dog-cart driven by a rather slightly-built but well set-up young man with a light moustache, bronzed skin, and brilliant blue eyes. He was good-looking, but if his features had been absolutely plain he could never have looked commonplace, for this was Lord Lester Leighton, son of the Earl of Kyneston, and twenty generations of unblemished descent had made him the aristocrat that he was.

Nitocris did not like pompous announcements by servants, and so she received her guests, who were all acquaintances or friends, in the great porch through which many a brilliant presence had passed, and had two maids waiting inside to see to the wants of the ladies, and their own coachman and a couple of grooms to attend to matters outside.

Merrill was made as happy as possible by a bright smile, a real hand-clasp instead of the usual Society paw-waggle, and instructions to go and make himself agreeable and useful. Brenda also received a hearty "shake" — Nitocris did not believe in kissing in public — and when the Professor and Mrs Huysman had gone in, she whispered:

"I suppose that's the Prince's brougham. You must wait here, dear, and do the introductions. You're responsible, you know."

Brenda assented with a nod and a smile, as the brougham drew up and the smart tiger jumped down and opened the door. The Prince got out, and was followed by Phadrig the Adept. As she looked at the two men, Nitocris felt as though a wave of cold air had suddenly enveloped her whole being — body and soul.

"Niti, this is our friend, Prince Oscar Oscarovitch, whom you have been kind enough to let me invite by proxy. Prince, this is Miss Nitocris Marmion."

Of course all the world knew of Oscar Oscarovitch, the modern Skobeleff, the lineal descendant of Ivan the Terrible, the crystal-brained, steel-willed man who was to be the saviour and regenerator of half-ruined, revolution-rent Russia, but this was the first time that Nitocris had met him in her present life. When she had returned his stately bow, she looked up and saw with a strange intuition, which somehow seemed half-reminiscent an almost perfect type of the primitive warrior through the disguise of his faultless twentieth-century attire. He was nearly two inches over six feet, but he was so exquisitely proportioned that he looked less than his height. His skin was fair and smooth, but tanned to an olive-brown. His forehead was of medium height, straight and square, with jet-black brows drawn almost straight across it above a pair of rather soft, dreamy eyes that were blue or black according to the mood of their possessor. His nose was strong and slightly curved, with delicately sensitive nostrils. A dark glossy moustache and beard trimmed *à la* Tsar, partly hid full, almost sensual lips and a powerful somewhat projecting chin.

As their eyes met the shiver of revulsion passed through her again. She hardly heard his murmured compliments, but her attention awoke when he turned to the man who was standing behind him, and said with a very graceful gesture of his left hand:

"Miss Marmion, this is the gentleman whom you have so graciously permitted me to bring to your house. This is Phadrig the Adept, as he is known in his own ancient land of Egypt, a worker of wonders which really are wonders, and not mere sleight-of-hand conjuring tricks. He has been good enough to accompany me in order to convince the learned of the West that the Immemorial East could still teach it something if it chose."

Nitocris bowed, and as she looked at the figure which now stood beside the Prince, she shivered again. She had a swift sense of standing in the presence of implacable enemies, and yet she had never seen these men before, and, for all she knew, she had not an enemy in the world. She was intensely relieved when Lord Lester Leighton came up and held out his hand, and she was able to ask the Prince and his companion to go through to the lawn.

No one would have recognised the shabby denizen of the grimy room in Candler's Court, Borough High Street, in the tall, dignified Eastern gentleman who walked with slow and stately step through the spacious old hall of "The Wilderness." He was clad in a light frock-coat suit of irreproachable cut and fit. The correctly-creased trousers met brightly-burnished, narrow-toed tan boots; a black-tasselled scarlet tarbush was set square on his high forehead, and the dark red tie under his two-ply collar just added the necessary touch of Oriental colour to his costume, and went excellently with the lighter red of the tarbush. It is hardly necessary to say that when he and the Prince went out on to the lawn, they were, as a Society paper report of the function would have put it, "the observed of all observers."

"I'm so glad you were able to be here in time for my little party, Lord Leighton," said Nitocris, when she had ended the welcoming of the other guests. "Dad will be delighted, too— —"

She stopped rather suddenly, remembering that Dad would have to tell his young friend the sad story of the mysterious loss of the Mummy; but another subject was uppermost in her mind just then, and, taking refuge in it, she went on quickly:

"Come along to the lawn. I want to introduce you to a very distinguished gentleman—and his wife and daughter. No less a person, my lord, than the great Professor Hoskins van Huysman!"

"What!" exclaimed Leighton, with a laugh that was almost boyish for such a serious and learned young man. "*The* Huysman: the Professor's most doughty antagonist in the arena of symbols and theorems? Oh, now that *is* good!"

"Yes; I think you will find him very interesting," replied Nitocris, hoping in her soul that he would find Brenda a great deal more interesting. "Come along, or Dad will be beginning to think that I am neglecting my duties, and I must be on quite my best behaviour to-day. We are favoured by the presence of another very celebrated celebrity to-day. That tall man who came in just before you was Prince Oscar Oscarovitch."

"Oh yes," he said lightly; "I recognised the brute."

"The brute? Dear me, that is rather severe. Then you know His Highness?" she asked in a low, almost eager, voice.

"There are not many men in the Near or Far East who have not some cause to know His Highness," he replied in a serious tone, tinged by the suspicion of a sneer. "He is about the finest specimen of the well-veneered savage that even Russia has produced for the last century. He is a brilliant scholar, statesman, and soldier; delightful among his equals—or those he chooses to consider so—charming to men, and, they say, almost irresistible to women; but to his opponents and his inferiors, a pitiless brute-beast without heart, or soul, or honour. A curious mixture: but that's the man."

"How awful!" murmured Nitocris. "Fancy a man like that being in such a position!"

But, although she did not understand why, she had heard his harshly-spoken words with a positive sense of relief. They exactly translated and crystallised her first inexplicable feelings of desperate aversion—almost of terror.

She led Leighton to a little group on the left side of the lawn, composed of the three Professors and the wives and daughters of two of them. As they approached them, Nitocris became sensible of a curious kind of nervousness. She did not know that by this commonplace action she was reuniting two links in a long-severed chain of destiny, but she had a dim consciousness that she was going to do something much more important than merely introducing two strangers to each other. She looked quite anxiously at

Brenda, who had turned towards them as they came near, and saw that, just for the fraction of a second, her eyes brightened, and a passing flush deepened the delicate colour in her cheeks. It was almost like a glance of recognition, and yet she had only heard his name two or three times, and certainly had never seen him before. Then she looked swiftly at Leighton. Yes, there was a flush under his tan and a new light in his eyes. When she had completed the introductions she looked away for a moment, and said in her soul:

"Thank goodness! If that is not a case of love at first sight, I shan't believe that there is any such thing, whatever the poets and romancers may say."

Yes, her womanly intuition was right as far as it reached; but she could not yet grasp the full meaning of the marvel which she had helped to bring about. With her father, she believed in the Doctrine of Re-Incarnation as the only one which affords a logical and entirely just solution of the bewildering puzzles and ghastly problems of human life as seen by the eyes of ignorance. She had grasped in its highest meaning the truth—that Man is really a living soul, living from eternity to eternity. An immortality with one end to it was to her an unthinkable proposition which could not possibly be true. For her, as for her father, Eternal Life and Eternal Justice were one. Where a man ended one life, from that point he began the next: for good or for evil, for ignorance or for knowledge. A life lived and ended in righteousness (not, of course, in the narrow theological sense of the term) began again in righteousness, and in evil meant inexorably a re-beginning in evil. That was Fate, because it was also immutable Justice. Man possessed the Divine gift of free will to use or abuse as he would, so far as his own life-conduct was concerned; but there was no evasion of the adamantine law of the survival and progress of the fittest, which, in the course of ages, infallibly proved to be the best. This, in a word, was why "some are born to honour and some to dishonour."

Yet she had still to fathom an even subtler mystery than this: the mystery of sexual love. Why should one man and one woman, out of all the teeming millions of humanity, be irresistibly attracted to each other by a force which none can analyse or define? Why should a woman, confronted with the choice between two men, one of whom possesses every apparent advantage over the other, yet feel her heart go out to that other, and impel her to follow him, even to the leaving of father and mother and home, and all else that has been dear to her? Why in the soul of every true man and woman is Love, when it comes, made Lord of all, and all in all? It is because Love is co-eternal with Life, and these two have loved, perchance wedded, many times before in other lives which they have lived together, and, with the succession of these lives, their love has grown stronger and purer, until "falling in love" is merely a recognition of lovers; unconscious, no doubt, to those who have not progressed far enough in wisdom, but none the less necessary and inevitable for that. [1]

Is it not from ignorance of this truth, or wilful denial of this law, that all the miseries of mismarriage come forth? Again the woman has the choice. She obeys the bidding of her own lust of wealth and comfort and social power, or she submits to the pressure of family influence, or the stress of poverty, and crushes—or thinks she does—the ages-old love out of her heart and marries the man she does not love, never has loved, and never can. She has defied the eternal Law of Selection. She has desecrated the sanctity of an immortal soul, and she has defiled the temple of her body. She has sold herself for a price in the market-place, and has become a prostitute endowed by law with a conventional respectability, and for this crime she pays the penalty of unsated heart-hunger. Instead of the fruits of Eden distilling their sweet juices into her blood, the apples of Gomorrah turn perpetually to ashes in her mouth. Often weariness and despair drive her to the brief intoxication of the anodyne of adultery, a further crime which is only the natural consequence of the first.

But it must not be thought that women are the only sexual criminals. There are male as well as female prostitutes made respectable by convention, and the debt-burdened man of title who marries to get gold to re-gild his tarnished coronet is the worst of these; for too often he drags an innocent but ignorant maiden down to his own vile level. Yet the chief criminal of all is not the individual, but the Society which not only encourages, but too often compels the crime. For this it also pays the penalty. The collective crime brings the collective curse, for, if human history proves anything, it proves that the Society which persistently denies the Law of Selection, and continually defiles the Altar of Love, in the end goes down through a foul welter of lust and greed and gluttony into the nethermost Pit of Destruction.

Nitocris had not learned this yet. It was not within the plan of Eternal Justice that her virgin soul, purified by the strenuous labour of many lives towards the Light, should yet be darkened by the shadow of such grim knowledge as this. It was enough for her now that she should be the ministering angel of Love and Light.

But at the same moment, standing on that smooth, shady lawn, there were also two incarnations of the destroying angels of Hate and Darkness, for even here, amidst this pleasant scene of seemingly innocent pleasure and laughter, the Eternal Conflict was being continued, as it is and must be, wherever man comes in contact with his kith and kind.

Soon after Nitocris and Brenda had joined the group, Phadrig approached the Prince, who happened for the moment to be standing alone at the bottom of the lawn, and said softly in Russian:

"Highness, my dream, as you are pleased to call it, has proved true. That is the Queen—she who was once the daughter of the great Rameses, Lady of the Upper and Lower Kingdoms."

"What?" laughed the Prince. "Miss Marmion, that lovely English girl, your old Egyptian Mummy re-vivified! Well, have it as you like. You are welcome to your dreams as long as you use your arts to help me to lay hands on the beautiful reality. I have seen many a fair woman, and thought myself in love with some of them, but by the beard of Ivan, I have never seen one like this. I tell you, Phadrig, that the moment my eyes looked for the first time into hers, only a few minutes ago, I knew that I had found my fate, and, having found it, I shall take very good care that I don't lose it. And you shall help me to keep it; I shall try every fair means first to make her my princess, for, whether she was once Queen of Egypt or not, she is worthy now to sit beside a sovereign on his throne—and it might be that I could some day give her such a place—but have her I will, if not as fairly-won wife and consort, then as stolen slave and plaything, to keep as long as my fancy lasts. And listen, Phadrig," he went on in a low tone, but with savage intensity. "Your life is mine, for I gave it back to you when the lifting of a finger would have sent you into what you would call another incarnation; and from this day forth you must devote it to this end until it is attained, one way or the other. I know you don't care for money as wealth, but in this world it is the right hand of power, and that you love. All that you need shall be yours for the asking in exchange for your faithful service. Are you content with the bargain?"

"No, Highness, that will not content me," replied Phadrig, in a voice that had no expression save unalterable resolve.

"What! Is not that enough for you, a penniless seller of curios?" said the Prince, with a sneer in his tone. "Then I will add to it the ready aid and unquestioning obedience of our secret police, here and in Europe. Will that satisfy you?"

"I do not need the help of your police, Highness," answered the Egyptian, in the same passionless accents. "They are skilful and brave, but they have not the Greater Knowledge. I could turn the wisest of them into a fool, and frighten the bravest out of his senses in a few minutes. Use them yourself, Highness, should it become necessary. They would be less than useless to me."

"Then what will satisfy you?" asked the Prince impatiently, but with no show of anger, for he knew the strange power of the man whose help he needed.

"I do not ask you to believe in the reality of what you call my dreams, Highness," replied Phadrig slowly, "but I do ask—nay, I require, as the price of my faithful service, your solemn promise in writing, signed and attested, that, if and when my dreams become realities, and your own hopes are fulfilled, the independence and sovereignty of the Ancient Land shall be restored; her temples and tombs and palaces shall be rebuilt; her ancient worship revived in my person, and the sceptre of Rameses replaced in the hand of Nitocris the Queen."

The Prince was silent for a few moments. To grant the seemingly extravagant demand meant to reduce the splendid dream and scheme of his life to cold, tangible writing, and to put into this man's hand the power to betray him. On the other hand, their aims were one, and only through him could Phadrig hope to realise his dreams. Of course they were only dreams; but he was faithful to them, and so he would be faithful to him. At the worst it would be easy to arrange a burglary, or, for the matter of that, a murder in Candler's Court, and that would make an end of the matter.

"Very well, Phadrig," he said at length. "It is settled. I will trust you, for it is necessary that we should trust each other. You shall have what you ask for within a week. Now I must go. I shall tell them that I have been arranging the exhibition of your powers which you are going to give them. It will be well to startle them sufficiently to shake their British beef-sense up into something like fear. Make them wonder, but, for the sake of our hostess, don't frighten them too much."

Phadrig only acknowledged his promise with a bow, and he turned away and joined the growing group in which Nitocris and Brenda were still the central objects of attraction.

[1] The Doctrine, of course, affords the same explanation of friendships between man and man, and woman and woman.

CHAPTER XI
THE MARVELS OF PHADRIG

The time, about an hour or so before tea, was occupied by the guests according to their varying tastes—in tennis, croquet, more or less good-natured gossip, and flirtations which may or may not have been serious.

Nitocris saw with growing cause for self-gratulation that Lord Leighton and Brenda were decidedly attracted towards each other. He, in spite of having received his gracious, but, as he well knew, final *congé* from Nitocris, still felt that he was not quite playing the game with himself; but for all that it was impossible for him not to see that the emotion, which was even now stirring in his heart, awakened by the first touch of Brenda's hand, and the first meeting of their eyes, was something very different from the tenderly respectful admiration, the real friendship, inevitably exalted by the magic of sex, which, as he saw now, he had innocently mistaken for love.

He managed quite adroitly to separate Brenda from the circle, and to lure her into a stroll about the outside grounds, during which he told her the history and traditions of "The Wilderness" not, of course, omitting the sad little tragedy of the Lady Alicia, all of which Miss Brenda listened to with an interest which was not, perhaps, wholly derived from the story itself. She had never yet met any one who was quite like this learned, much-travelled, quiet-spoken young aristocrat. On her father's side she was descended from one of the oldest Knickerbocker families in the State of New York and her aristocracy responded instinctively to his, and formed a first bond between them.

It need hardly be said that her beauty and her prospective wealth, to say nothing of the bright, mental, and intellectual atmosphere in which she seemed to live and move, had attracted to her many men whom she had inspired with a very genuine desire to link their lives with hers. She was only twenty-two, but she had already refused more than one coronet of respectable dignity, and so far her heart had remained as virgin as it was when she had admired herself in her first long skirt. But now, for the first time in her life, she began to feel a strange disquietude in the presence of a man, and a man, too, whom she had not known for an hour. Nitocris had, happily, told her nothing of what had passed between Lord Leighton and herself, and so the pleasant element in her disquietude was entirely unalloyed.

Her father was already too deeply engrossed in learned converse with his brother professors to take any notice of the great fact which was beginning to get itself accomplished; but her mother's instinct instantly noticed the subtle change that had come over her daughter, and she saw it with anything but displeasure. All sensible mothers of beautiful daughters are discreetly sanguine. She was far too wise in her generation not to have agreed with Brenda's decision in certain former cases. The idea of her daughter's beauty and her father's millions being bartered for mere rank and social power, however splendid, was utterly repugnant to her. She had married for love, and she wanted Brenda to do the same, whoever the chosen man might be, provided always that he was a man—and in this regard there could be no doubt about Lord Lester Leighton; so as they walked away she said to Nitocris with a confidence which was almost girlish:

"His Lordship is just delightful—now, isn't he, Miss Marmion? Just the sort that you seem to raise over here, and nowhere else. Tells you that you have to take him for a gentleman and nothing else in the first three words he says to you—and Brenda seems to like him. I never saw her go off with a man like that on such short notice, for Brenda's pretty proud and cold with men, for all her nice ways and high spirits."

"You would have to search a long time, Mrs van Huysman," replied Nitocris very demurely, "before you found a better type of the real English gentleman than Lord Leighton. His family is one of the oldest in the country, and, unlike too many of our noble families, the Kynestons have no bar-sinister on their escutcheon."

"I guess you're getting a little beyond me there, Miss Marmion. I don't think I ever heard of a—what is it?—a bar-sinister, before. What might it be?"

Nitocris flushed very faintly as she replied:

"I think I can explain it best, Mrs van Huysman, by saying that it means that Lord Leighton's ancestors have preserved their honour unstained through many generations. Of course, you know that some of our so-called noble families in England spring from anything but a noble origin. There are not a few English dukes and earls who would find it rather awkward to introduce their great-great-grandmothers to their present circle of friends."

"I should think they would, from what I have read of them, the shameless creatures!" said Mrs van Huysman, with a sniff of real republican virtue.

Then the Prince joined them, and the conversation was promptly switched off on to another line of interest.

Tea was served on the Old Lawn under the shade of the great cedars, which made its greatest adornment; and when everybody had had what he or she wanted, and the men had lit their cigarettes—and the Professors, by special permission, their pipes—Nitocris looked across a couple of tables at Oscarovitch, whom she had so far managed most adroitly to keep at an endurable distance, and said:

"Now, Prince, if your friend the Adept is in the mood to astonish us with his wonders, perhaps you will be good enough to tell him that we are all ready and willing to be startled—only I hope that he will be merciful to our ignorance and not frighten us too much."

"I can assure you, Miss Marmion, that my good friend from Egypt will be discretion itself," replied the Prince, with a look and a courtly gesture that inspired Commander Merrill with an almost passionate longing to take him down one of the quiet paths under the beeches for a ten minutes' interlude. "I can promise that he will show you some marvels which even your learned and distinguished father and his *confrères* may find difficult of explanation: but it shall all be white magic. I understand that your real adept considers the black variety as what you call bad form."

As the company rose and went in little groups towards the tennis-lawn, where Phadrig had elected to display his powers, the three Professors instinctively joined each other in a small phalanx of scepticism. If there was any trick or deception to be discovered all looked to them to do it, and they were almost gleefully aware of their responsibility. Figuratively speaking, they each wore the scalps of many spiritualistic mediums, and both Professor van Huysman and Professor Hartley sensed a possible addition to their belts of scientific wampum which would not be the least of their trophies. It had been agreed to by Phadrig, with a quiet scorn, that they were to take any measures they liked to detect him in any practice that would convict him of being merely a conjurer; and they had accepted the permission with that whole-souled devotion to truth which excludes all idea of pity from the really scientific mind. Franklin Marmion was naturally in a very different frame of mind, although, from reasons of high policy, he assumed a similar mask of almost scornful scepticism; but for all that he was by far the most anxious man in the company.

At the request of their hostess the guests arranged themselves sitting and standing in a spacious circle on the tennis-lawn; and when this was, formed, Phadrig, whose isolation so far from the rest of the company had been satisfactorily explained by the Prince, walked slowly into the middle of it, and, after a quick, keen glance round him—a glance which rested for just a moment or so on Professor Marmion and his *confrères*, and then on Nitocris, who was sitting beside Brenda attended by Lord Leighton and Merrill—he said in a low but clear and far-reaching voice, and in perfect English:

"Ladies and gentlemen, I have come to the house of the learned Professor Marmion at the request of my very good friend and patron, His Highness Prince Oscar Oscarovitch, to give you a little display of what I may call white magic. But before I begin I must ask you to accept my word of honour as a humble student of the mysteries of what, for want of a better word, we call Nature, that I am not in any sense a conjurer, by which I mean one who performs apparent marvels by merely deceiving your senses.

"What I am going to show you, you really will see. My marvels, if you please to think them such, will be realities, not illusions; and I shall be pleased if you will take every means to satisfy yourselves that they are so. I say this with all the more pleasure because I know that there are present three gentlemen of great eminence in the world of science, and if they are not able to detect me in anything approaching trickery, I think you will take their word for it that I am not deceiving you.

"In order that there may not be the smallest possible chance of error, I will ask Professors Marmion, Hartley, and Van Huysman to come and stand near to me, so that they may be satisfied that I make use of none of the mere conjurer's apparatus. I shall use nothing but the knowledge, and therefore the power, to which it has been my privilege to attain."

Phadrig spoke with all the calm confidence of perfect self-reliance, and therefore his words were not wanting in effect on his audience, critical and sceptical as it was.

"I reckon that's a challenge we can't very well afford to let go," said Professor van Huysman, with a keen look at his two brother scientists. "Of course he's just a trick-merchant, but they're so mighty clever nowadays, especially these fellows from the gorgeous East, that you've got to keep your eyes wide open all the time they've got the platform."

"Certainly," said Professor Hartley, as they moved out from the circle; "it must be trickery of some sort, and we shall be doing a public service by exposing it. What do you think, Marmion? I hope you won't mind the exposure taking place in your own garden and among your own guests?"

"Not a bit, my dear Hartley," replied Franklin Marmion with a smile, which was quite lost upon his absolutely materialistic friends. "We have, as Van Huysman says, received a direct challenge. We should be most unworthy servants of our great Mistress if we did not take it up. Personally, I mean to find out everything that I can."

"And, gentlemen," laughed the Prince, who had been standing with them and now moved away towards Nitocris, "I sincerely hope that what you find out will be worth the learning."

"He's a big man, that," said Professor van Huysman, when he was out of earshot, "but he's not the sort I'd have much use for. I wonder why those people who are on the war-path in his country ever let him out of it alive?"

In accordance with Phadrig's request, they made a triangle of which he was the central point. Without any formula of introduction, he said rather abruptly:

"Professor van Huysman, will you oblige me by taking a croquet ball and holding it in your hand as tightly as you can?"

Brenda ran out of the circle and gave him one. He took it and gripped it in a fist that looked made to hold things. Phadrig glanced at the ball, and said quietly:

"Follow me!"

Then he turned away, and, in spite of all the Professor's efforts to hold it, the ball somehow slipped through his fingers and fell on to the lawn. Then, to the utter amazement of every one, except Franklin Marmion, it rolled towards the Adept and followed him at a distance of about three yards as he walked round the circle of spectators. He did not even look at it. When he had made the round, he took his place in the Triangle of Science, and the ball stopped at his feet.

"It is now released, Professor," he said to Van Huysman. "You may take it away, if you wish."

There was something in the saying of the last sentence that nettled him. He had seen all, or nearly all, the physical laws, which were to him as the Credo is to a Catholic or the Profession of Faith to a Moslem, openly and shamelessly outraged, defied, and set at nought. To say he was angry would be to give a very inadequate idea of his feelings, because he, the greatest exposer of Spiritualism, Dowieism, and Christian Scientism in the United

States, was not only angry, but—for the time being only, as he hoped—utterly bewildered. It was too much, as he would have put it, to take lying down, and so, greatly daring, he took a couple of strides towards Phadrig, and said with a snarl in his voice:

"I guess you mean really if *you* wish, Mr Miracle-Worker. It was mighty clever, however you did it, but you haven't got me to believe that physical laws are frauds yet. You want me to pick that ball up?"

"Certainly, Professor—if you can—now," replied Phadrig, with a little twitch of his lips which might have been a smile, or something else.

Hoskins van Huysman was a strong man, and he knew it. Not very many years before, he had been able to shoulder a sack of flour and take it away at a run, and now he could bend a poker across his shoulders without much trouble. He stooped down and gripped the ball, expecting, of course, to lift it quite easily. It didn't move. He put more force into his arms and tried again. For "all the move he got on it," as he said afterwards, it might have weighed a ton. It was ridiculous, but it was a fact. In spite of all his pulling and straining, the ball remained where it was as though it had been rooted in the foundations of the world. He was wise enough to know when he was beaten, so he let go, and when he pulled himself up, somewhat flushed after his exertions, he said:

"Well, Mister Phadrig, I don't know how you do it, but I've got to confess that it lets me out. I'm beaten. If you can make the law of gravitation do what you want, you're a lot bigger man in physics than I am."

He turned and went back to his place, looking, as his daughter whispered to Nitocris, "pretty well shaken up." The Prince caught Phadrig's eye for an instant, and said:

"Miss Marmion, will you confound the wisdom of the wise and bring the ball here?"

It was not the words but the challenge in them that impelled her to rise from her chair, aided by Merrill's hand, and not the one that the Prince held out, and walk across the lawn towards Phadrig. She took no notice of him. She just stooped and picked up the ball and carried it back to her chair. She tossed it down on the grass, and sat down again without a word, quaking with many inward emotions, but outwardly as calm as ever. What Professor van Huysman said to himself when he saw this will be better left to himself.

It might have been expected that the miracle, or at least the extraordinary defiance of physical law which had been accomplished by Phadrig, would have produced something like consternation among the bulk of the spectators. It did nothing of the sort. They were, perhaps, above the ordinary level of Society intellect in London; but they only saw something wonderful in what had been done. Nothing would have persuaded them that it was not the result of such skill as produced the marvels of the Egyptian Hall, simply because they were not capable of grasping its inner significance. Could they have done that, the panic which Professor Marmion was beginning to fear would probably have broken the party up in somewhat unpleasant fashion. As it was they contented themselves with saying: "How exceedingly clever!" "He must be quite a remarkable man!" "I wonder we've never heard of him before!" "He must make a great deal of money!" "I wonder if I could persuade the dear Prince—what a charming man he is!—to bring him to my next At Home day?" and so on, perfectly ignorant, as it was well they should be, that they had witnessed a real conquest of Knowledge over Force.

Phadrig, who seemed to be the least interested person on the lawn, looked about him, and said as quietly as before:

"I should be very much obliged if the best tennis player in the company will do me the honour to have a game with me."

Now, it so happened that Brenda, in addition to her other athletic honours, had recently won the Ladies' Tennis Tournament at Washington, which carried with it the Championship of the State for the year, and so this challenge appealed both to her pride in the game and her spirit of adventure. She looked round at Nitocris, and said:

"I've half a mind to try, Niti. I suppose he won't strike me with lightning or send me down through the earth if I happen to beat him. Shall I?"

"Yes, do," replied her hostess, with a suspicion of mischief in her voice; "those dear Professors of ours are puzzling so delightfully over the first miracle, or whatever it was, that I *do* want to see them worried a little more. It will be a wholesome chastening for the overweening pride of knowledge."

"Very well," laughed Brenda, rising and dropping a light cloak from her shoulders. "It's the first time I've had the honour of playing against a magician, mind, so you mustn't be too hard on me if I lose."

Lord Leighton fetched her racquet and one for Phadrig, and they went together towards the tennis-court in which he was standing. The three Professors left their places and stood at one end of the net, Messrs Hartley and Van Huysman indulging in audible growls of baffled scepticism, and Franklin Marmion silently observant, divided between interest and amusement. He could not help imagining what would happen if he were to stand in the middle of the circle and remove himself to the Higher Plane, and then go round shaking hands and saying, "Good afternoon."

Brenda acknowledged Phadrig's bow with a gracious nod as she took her place. Then Lord Leighton handed the other racquet to the Adept. To his astonishment he declined it with another bow, saying:

"I thank you, my lord, but I do not need it."

"What!" exclaimed the other, with a frank stare of astonishment. "Excuse me, but tennis without a racquet, you know—are you going to play with your hands?"

"To some extent, yes, my lord," replied Phadrig, as he took his place. "Will you ask Miss van Huysman if she will be kind enough to serve?"

Brenda would. Phadrig stood on the middle line between the two courts with his hands folded in front of him. She certainly felt a little nervous, but she knew her skill, and she sent a scorcher of an undercut skimming across the net. The ball stopped dead. Phadrig gave a flick with his right forefinger, and it hopped back over the net and ran swiftly along the ground to Brenda's feet. She flushed as she picked it up and changed courts. Then she raised her racquet and sent a really vicious slasher into the opposite court. Phadrig, without moving, raised his hand at the same moment. The ball, hard as it had been driven, stopped in mid-air over the net, hung there for a moment, then dropped on Brenda's side and rolled to her feet again. She picked it up, walked to the net with it in her hand, and said quite good-humouredly:

"I think you're a bit too smart for me, Mr Phadrig. I can't pretend to play against a gentleman who can suspend the law of gravitation just to win a game of tennis."

"I did not do it to win the game, Miss van Huysman," he replied with a gentle smile; "I only desired to amuse you and the other guests of Professor Marmion. Now, it may be that some excellent but ignorant people here may think that that ball is bewitched, as they would call it, so if you will give it to me, I will send it out of reach."

She handed him the ball, wondering what was going to happen next. He took it and put it on the thumb of his right hand as one does with a coin when tossing. He flicked it into the air, and, to the amazement of every one, saving always Franklin Marmion, it rose slowly up to the cloudless sky, followed by the gaze of a hundred eyes, and vanished. Then he bowed again to Brenda, and said in the most commonplace tone:

"It is out of harm's way now. Thank you once more for your condescension."

"But how did it go up like that?" asked Brenda, looking him frankly and somewhat defiantly in the eyes.

"That, Miss Huysman," he replied with perfect gravity, "was only a demonstration of what Spiritualists and Theosophists are accustomed to call levitation. It is only a matter of reversing the force of gravity."

"Is that all?" laughed Brenda, as she turned away. "You talk of it as though it were a matter of turning a paper bag inside out."

"The one is as easy as the other," he smiled. "It is only a question of knowing how to do it."

She walked back to her chair very much mystified, and, for the first time in her so far triumphal journey through the interlude between the eternities which we call life, a trifle humiliated: but that fact, of course, she kept to herself. As she dropped back in her chair, she said to Lord Leighton:

"That was pretty wonderful, wasn't it? I'm quite certain that there's no trickery about it. What he did, he really did do."

"I don't pretend to be able to explain it," he replied, "but for all that I've seen very much the same sort of thing done by the fakirs in India, and I think it's generally admitted that that is either a matter of trickery or hypnotism. They make you believe you see what you really don't see at all."

"That's about it," said Merrill, with a short laugh, "Of course no one who knows anything about the East will deny that hypnotism is a fact, although I must say that these same fakirs have tried it with me more than once and found me a quite hopeless subject."

Even as though he had heard him, Phadrig came towards them at the moment, and said in his polite, impersonal tone:

"Commander Merrill, I am going to try one or two experiments now which I should like to have very closely watched. I know that there is no keener observer in the world than the skilled British naval officer. May I ask for your assistance?"

There was something in his tone which made it quite impossible to refuse, so he replied:

"You have shown us a good many wonders already, Mr Phadrig, and unless you've hypnotised the whole of us, I haven't a notion how you have done it; but if I can find you out I will."

"That is exactly what I wish, sir," said Phadrig, as he bowed to the ladies and went back to the centre of the circle. Merrill followed him, and, with the three Professors, formed a square about him.

Phadrig, turning slowly round so that his voice might reach all his audience, said:

"Ladies and gentlemen, you have all heard of or seen the strange performances of the Indian fakirs: the growing of the mango plant, the so-called basket trick, and the throwing into the air of a rope up which the performer climbs from view of the spectators. I am not going to say whether those are tricks or not. Their knowledge may be different from mine, therefore I do not question it. I only propose to show you the same kind of performance without the use of any coverings or concealment, and leave you and these four gentlemen to discover any deception on my part if you can. I will begin by giving you a new version of the mango trick, if trick it is, with variations. Professor Marmion, would you have the goodness to ask one of the young ladies to bring me one of those beautiful white roses of yours?"

Franklin Marmion was on the point of saying: "I'll bring you one myself, and see what you can do with it," but he was a sportsman in his way, and, seeing that his guests were so far not all inclined to be frightened at what they had seen, he refrained from spoiling the "entertainment," as they evidently took it to be, and so he asked his daughter to go and get one of her nicest Marèchal Niels.

She rose from her chair and went to her favourite tree; Merrill followed her with a ready penknife. They came back with a fine half-blown rose on a leafy twig about nine inches long. As she held it out to Phadrig he declined it with a bow and a wave of his hand, saying:

"I thank you, Miss Marmion, but it will be better for me not to touch it. Some one might think that I had bewitched it in some way; will you be kind enough to give it to Commander Merrill and ask him to put the stem into the turf: about two inches down, please."

She handed the rose to Merrill, and as he took it their eyes met for an instant, and she flushed ever so slightly. He, with many unspoken thoughts, knelt down, made a little hole in the turf with his knife, and planted the rose. When he stood up again Phadrig went on in the same quiet impersonal voice:

"Now, ladies and gentlemen, you know that this rose is of a pale cream colour slightly tinted with red. It shall now grow into a tree bearing both red and white roses. It will not be necessary for me to touch it."

This somehow appealed more closely to such imagination as the majority of the spectators possessed. They had regarded the other marvels they had seen merely as bewilderingly clever examples of legerdemain: but for a man to make a single sprig of rose grow into a tree bearing both red and white roses without even touching it meant something quite unbelievable—until they had seen it. Instinctively the circle narrowed, and Phadrig noting this, said:

"Pray, come as close as you like, ladies and gentlemen, as long as you do not pass my guardians, for they have undertaken that you shall not be deceived."

The result was that a smaller circle was formed round the square, at the angles of which stood Merrill and the three men of science. Phadrig stood at one side facing the east. Then he spread his hands out above the rose, and said slowly:

"Earth feeds, sun warms, and air refreshes: wherefore grow, rose, that the power of the Greater Knowledge may be manifested, and that those who believed not before may now see and believe."

He raised his hands with a spreading movement and, to the utter amazement of every one except Franklin Marmion, who now saw that this man certainly had approached to within measurable distance of the borderland which he had himself so lately crossed—wherefore in his eyes there was nothing at all marvellous in anything he had done—the leaves on the sprig grew rapidly out into branches as the main stem increased in height and thickness, red and white buds appeared under the leaves and swelled out into full blooms with a rapidity that would have been quite incredible if a hundred keen eyes had not been watching the marvel so closely; and within ten minutes a fine rose-bush, some three feet high, loaded with red and white and creamy blossoms, stood where Merrill had planted the sprig.

After the first gasps of astonishment there arose quite a chorus of requests from the younger members of Phadrig's audience for a rose to keep in memory of the marvel they had seen; but he shook his head, and said with a smile of deprecation:

"I regret that it is not possible for me to grant what you ask. For your own sakes I cannot do it. If I gave you those roses they would never fade, and it might be that those who possessed them would never die. Far be it from me to curse you with such a terrible gift as immortality on earth."

The gravely, almost sadly spoken words fell upon his hearer's ears like so many snowflakes. Instinctively they shrank back from the beautiful bush as though it had been the fabled Upas. They had begun to fear now for the first time. But there was one among them, a young fellow of twenty-two, named Martin Caine, who was already known as one of the most daring and far-sighted of the rising generation of chemical investigators, to whom the prospect of an endless life devoted to his darling science was anything but a curse. Intoxicated for the moment by what he had seen, he sprang forward, exclaiming:

"I'll risk the curse if I can have the life!"

As his hand touched one of the roses, Phadrig's darted out and caught his wrist. He was a powerful youth, but the instant Phadrig's hand gripped him he stopped, as though he had been suddenly stricken by paralysis. He turned a white, scared face with fear-dilated eyes upward, and said in a half-choked voice:

"What's the matter? If what you say's true, give me eternal life, and I'll give it to Science."

"My young friend," said Phadrig, with a slow shake of his head, "you are grievously mistaken. You have eternal life already. You may kill your body, or it may die of age or disease, but the life of your soul is not yours to take or keep. Only the High Gods can dispose of that. Who am I that I should abet you in defying their decrees? Here is my refusal of your mad request."

He plucked the rose which Caine had touched, held it to his lips and breathed on it. The next instant the withered leaves fell to the ground, and lay there dry and shrivelled. The stalk was brown and dry. As he released Caine's wrist he dropped the stalk in the middle of the bush, and said in a loud tone:

"As thou hast lived, die—as all things must which shall live again."

As quickly as the rose-bush had grown and flowered so quickly, it withered and died. In a few moments there was nothing left of it but a few dry sticks lying in a little heap of dust.

The circle suddenly widened out as the people shrank back, every face showing, not only wonder now, but actual fear; and now Franklin Marmion felt that Phadrig had been allowed to go as far as a due consideration for the sanity of his guests would permit. The other two Professors were disputing in low, anxious tones, as if even their scepticism was shaken at last: Martin Caine had drifted away through the opening press to hide his terror and chagrin. The Adept stood impassively triumphant beside the poor relics of the rose-bush, but obviously enjoying the consternation that he had produced—for now the lust of power which ever attends upon imperfect knowledge had taken hold of him, and he was devising yet another marvel for their bewilderment. But before he had arrived at his decision, something else happened which was quite outside his programme.

The Prince broke the chilly silence by saying to Nitocris in a tone loud enough for every one to hear:

"I hope, Miss Marmion, that I have justified my intrusion by the skill which my friend Phadrig has displayed for the entertainment of your guests?"

She turned and looked at him, and, as their glances met, he saw a change come over her. Her eyes grew darker: her features acquired an almost stony rigidity utterly strange to her. His eyelids lifted quickly, and he shrank back from her as a man might do who had seen the wraith of one long dead, but once well known.

"Nitocris!" he murmured in Russian. "Phadrig was right: it is the Queen!"

She swept past him—Oscar Oscarovitch, the man who aspired to the throne of the Eastern Empire of Europe—as though he had been one of his own slaves in the old days, and faced Phadrig.

"It is enough, Anemen-Ha that was. Hast thou not learned wisdom yet, after so many lives? Is the inmost chamber of thy soul still closed in rebellion against the precepts of the High Gods? No more of thy poor little mummeries for the deception of the ignorant! Go, and without further display of the weakness which thou hast presumptuously mistaken for strength. The Queen commands—go!"

Only Phadrig and Franklin Marmion saw that it was not Nitocris, the daughter of the English man of science, but the daughter of the great Rameses who stood there crowned and robed as Queen of the Two Kingdoms.

Phadrig raised the palms of his hands to his forehead, bowed before her, and murmured:

"The Queen has but to speak to be obeyed! It is even as I feared. But the Prince——"

"I who was and am, know what thou wouldst say. Go, or——"

"Royal Egypt, I go! But as thou art mighty, have mercy, and make the manner of my going easy."

Nitocris turned away with a gesture of utter contempt, walked slowly towards her father, and said in English:

"Dad, I think our friend the Adept is a little tired after his wonder-working. I dare say most of us would be if we could do what he has been doing. He seems quite exhausted. I think you had better ask the Prince to let his coachman take him home."

Oscar Oscarovitch's soul was in a tumult of bewilderment, but his almost perfect training made it possible for him to say as quietly as though he had been taking leave of his hostess at a reception in London:

"Miss Marmion, we must thank you for your great consideration. As you say, our friend is undoubtedly fatigued, and, as I have an appointment at the Embassy this evening, I will ask you to allow me to take my leave as well."

With a comprehensive bow of farewell to the company, and a somewhat limp handshake with Professor Marmion and his daughter, he put his arm through that of his defeated and humiliated accomplice, and led him away through an opening which the still dazed spectators instinctively made for them.

CHAPTER XII
CONTROVERSY AND CONFIDENCES

After this incident, the guests melted away, singly and by pairs and families, thanking Nitocris and her father with much *empressement* for "the delightful afternoon," and "the extraordinary entertainment which they had so much enjoyed," and many regrets that "the poor Adept, who really was so very clever and had mystified them all so delightfully," had overdone himself and got ill, and so on, and so on, through the endless repetitions and variations usual on such occasions.

A small party, including the Hartleys, the Van Huysmans, Merrill, and Lord Leighton, had been asked to stay to dinner, but it happened that they had a conversazione already included in the day's programme, and so they took their departure soon after the others, the Professor, it must be confessed, in a somewhat morose frame of mind. Like all men of similar mental constitution, he hated to be mystified, and now, for the first time in his long career of investigation into apparently abstruse phenomena, he had been absolutely stumped by this perfect-mannered, quiet-spoken gentleman from the East who performed wonders in broad daylight, on a plot of grass amidst a crowd of people, and did not deign to even touch the things he worked his miracles with. If he had only used some sort of apparatus, or condescended to some concealment, after the manner of others of this kind, there might have been a chance of finding a means of exposure; but the whole performance had been so transparently open and aboveboard that Professor Marcus Hartley, D.Sc., M.A., F.R.S., etc., etc., felt that, as a consistent materialist, he had not been given a fair chance. Still, he did not despair; and by the time he got back into his own den he had resolved that when it did come, as of course it must do sooner or later, the exposure of Phadrig the Adept and the vindication of Natural Law should be complete and final.

A discussion of the same marvels naturally bulked largely in the conversation during dinner at "The Wilderness." Mrs van Huysman did not contribute much wisdom to it beyond the assertion of her conviction that such things were wicked and should be stopped by law, at which her daughter was sufficiently unfilial to draw a diverting picture of a stalwart policeman trying to arrest an elusive adept who could probably make himself invisible at will, or call to his aid fire-breathing dragons, just as easily as he could make a tennis ball evaporate into thin air, or grow lovely witch-roses and wither them to ashes with a breath.

"I do think it was a bit mean of him not to let that poor young man have one of them, if he was willing to take the risk. Especially as he just wanted to go on working for Science for ever. Fancy what a single man might do if he could just keep right on with his life-work for, say, a thousand years without having to stop it to die and be born again, according to Niti's pet theory. What couldn't a man like that do for human knowledge!"

"Would you have had one of those roses, Brenda, if the Prince's miracle-worker had offered you one?" asked Nitocris, smiling, but still with a decided note of seriousness in her tone.

"I?" laughed Brenda, leaning back in her chair. "Sakes, no, child! I've had a pretty good time so far, and I hope it won't be over just yet; but, after all, there must be a limit even to the combinations of human life, and a time would have to come when you'd just be doing the same old things over and over again. And, besides that, think of the horror of living on and on and seeing every one you loved—husband and wife, and children and grandchildren—grow old and die, and leave you alone in a world of strangers. No; life's a good thing if you only have fair play in the world; but so is death when you've lived your life. It's only like going to bed, after all. Eternal life would be like a day with no night to it, and that, I guess, would get a bit monotonous after a century or two. What do you think, Professor?"

"My dear Miss van Huysman," replied her host with one of his rare but eloquent smiles, "since I began to study the question with anything like enlightenment, I have not been able to look upon what we call life, by which I mean existence in this or some other world, as anything but eternal. In its manifestations to our senses it is, I admit, merely transitory, a brief span of time between two other states which, for want of a better word, we may call two eternities; but I must confess that, to me, a human existence beginning with the cradle and ending with the grave is merely a more or less tragic riddle without an answer: in other words, a meaningless absurdity. I find it quite impossible to conceive any deity or presiding genius of the universe who could be guilty of such a colossally useless tragedy as human life would be under those circumstances."

"I can't see it, my dear Marmion," said Brenda's father a trifle gruffly, for he had not yet quite recovered from the disquieting experiences of the afternoon. "What does it matter whether we live again or not as long as we live cleanly and do our work honestly while we are alive? Surely if we leave this world a little bit better, a little bit richer in knowledge, than we find it, these poor little lives of ours, such as they are, and that's not much—will not have been lived in vain. Of course, as you know, I'm just a common, low-down materialist who can't rise to the poetry of things as you can with this gorgeous theory of re-incarnation of yours.

"I should very much like to believe it if I could, as I once said to an eminent revivalist on the war-path in the States; but the trouble with a man who is honest with himself is that he can no more make himself believe what doesn't seem true to him than he can make himself hungry when he isn't. All the horrible history of religious persecution is just the story of a lot of bigots in power trying to force helpless people to do what they couldn't do honestly. The awful part of the business is that they were most likely all wrong, and didn't know it."

"But, at least, Professor, I hope you are able to give them credit for honest intentions, however mistaken they might have been?" interposed Merrill, who was the son of a country parson and had so far preserved his simple faith intact. It may be remarked here, that Nitocris was well aware of this, and loved her strong-souled sailor all the better for it. Franklin Marmion did not, but then he thought any creed good enough for "a mere fighting man."

"There were schemers and scoundrels among them on both sides, sir," replied the American quietly. "The temptation was too big; but I am quite willing to allow that the majority of them, even the Inquisitors, were honest zealots who really did think it right to produce any amount of suffering and misery here on earth in order to get matters straightened out, as they thought, hereafter. Charles V. was the most enlightened monarch of his age and the worst persecutor, and Torquemada, away from his religion, was as kind-hearted a man as ever lived. Calvin was a good man, but he watched Servetus burn, and our own Pilgrim Fathers on the other side were just about as hard men as any when it came to arguing out a religious question with whips and pillories and thumbscrews, and the like. I don't want to offend any one's sentiment or question any one's faith. To each man the belief that satisfies him, but personally I have no use for a religion that can't get itself believed without persecution."

"I quite agree with you there, Professor," replied Merrill, who felt a little chilled by the perfect aloofness with which the other spoke, and was

The Mummy And Miss Nitocris A Phantasy Of The Fourth Dimension | 89 wondering what his dear old father, living his quiet, saintly life among the Derbyshire dales, would have thought of such cold-blooded heresy. "I have always looked upon that sort of brutal intolerance as a form of religious mania—sincere, but still mania, and the story of it is the most awful chapter in human history——"

"Except, perhaps, the story of war," interrupted Professor Marmion, with a snap in his voice. Monomania, more or less harmless, is a not infrequent affliction of very high intelligences, and a quite unreasoning hatred of war was his, although within the last few days he had come to suspect disquieting misgivings on the subject, possibly in consequence of the higher knowledge to which he was attaining.

"My dear sir," replied Merrill quite good-humouredly, and not at all sorry for the diversion, "I am glad to say that I agree with you also. No man who has not actually fought can have any just idea of the appalling abominations of war, and I am sure that no men hate it more devotedly than those who have to fight. But we have to take the world as it is, and not as we would like it to be; and as long as we have people in it who want to set it on fire for their own brutally selfish purposes, we shall have to keep the fire-extinguishers in good order."

In obedience to an appealing glance from his daughter, the Professor did not reply. His opponent in the bloodless arena of Science saved him by interrupting:

"Yes, sir. I differ from my friend Marmion on a good many points, and that's one of them. You have the honour to serve in the biggest fire-extinguishing institution on earth. It was the British Navy that put out Napoleon's bonfire that he was making of the world: you kept the ring round us and Spain, and round Russia and Japan, and you've saved more conflagrations than half a dozen Noah's floods would put out. That's why the Kaiser and his tin-hatted firebrands have such a healthy dislike for you. They'd have had the world on fire years ago if they hadn't had to worry about you."

"I think you must admit, Professor Marmion," said Lord Leighton, who had so far been busy with his own new thoughts and the contemplation of the inspirer of them, "that it is people like these on whom the real guilt of the crime of war rests. Now that the pressure of the bear's paw is removed, Germany is the danger-spot of the world. The Maroocan business proved that pretty clearly; and nothing but our friendship with America and France and Japan, and the ability to strike hard and instantly at sea, saved Europe, and perhaps the world, from something like a repetition of the Napoleonic wars."

"With Mister William Hohenzollern a Napoleon," added Professor van Huysman, with a half-suppressed snort. "It seems to me as though that gentleman had been spreading himself round Europe as German War-Lord so long that he's getting tired of playing at it, and 's just spoiling for a real fight."

"That is very possible," said Merrill; "but happily he has responsibilities, and even the German war party would not follow him as far as he would like to go, to say nothing of the Liberals and the Socialists. Personally, I must say that I think we have had a much more dangerous person, as far as the peace of the world is concerned, on the lawn of 'The Wilderness' this afternoon."

"Of course you mean that hateful Russian Prince who brought that equally hateful Adept, as he calls himself, with him," said Nitocris, with an unwonted harshness that made every one look up.

"Oh, Niti," exclaimed Brenda, "and I asked you to let me bring him!"

"I'm very sorry, dear," she replied quietly, but with a smile of reassurance. "It was not your fault, of course. He may have been very nice to you, but I am obliged to say that the first moment I looked at him I was possessed by some inexplicable feeling of dislike, and even fear, although I certainly never hated or feared any one before. If I had met him before I got your note, I really think I should have asked you to spare us the honour. It seemed to me as though there was something uncanny about the man. It was very curious."

Her father looked up at her for a moment, wondering what would happen if he were to explain the mysterious antipathy there and then. The little theological discussion would look very small after such a revelation as that. But he, too, had had a revelation which the somewhat desultory conversation had done something to press home upon him. He had seen the advent of the Queen, and heard what she had said to Phadrig with other eyes and ears than his guests had done, for to them it had only been Nitocris who had gone to him and said a few inaudible words, which they had taken as a request for the conclusion of his "performance."

He had seen back through the mists of many centuries and recognised them as they had been, and he had learned that Oscarovitch the Russian had now entered the circle of the Queen's, and therefore his own, influence. A sudden anxiety for the safety of his darling Niti had awakened in his heart. He had seen the lust for possession flame in the man's eyes, and now that he knew who he was—and had been—he determined that whatever other adventurer might set the world aflame, the Modern Skobeleff should not do it if he and his Royal ally on the Higher Plane could prevent it. His coming had been a curious coincidence, possibly a consequence of obscure causes; but, for some reason or other, he felt himself beginning to look with a more favourable eye on Commander Mark Merrill—perhaps because he was the impersonation of uncompromising hostility to everything that Oscarovitch represented.

Dinner had come to an end now, and so Nitocris took advantage of ending a conversation which bade fair to become somewhat awkward. She glanced round the table and rose, saying:

"Don't you think we've had polemics enough for one little dinner, Dad? There's a lovely moon, so we'll have our coffee on the verandah, and you and Mr van Huysman can settle the affairs of the universe comfortably over your pipes. Give Lord Leighton and Mr Merrill something to smoke, and we will join you when we have got some wraps."

When they got back from Nitocris's rooms Mrs van Huysman elected to take her coffee in a big, deep-seated armchair by the drawing-room window. She said that she had felt the sun a little, and might possibly indulge in forty winks—which she did within a few minutes of getting comfortably arranged in it. Then Nitocris took Brenda by the arm and walked her half-way down the lawn.

"I want to take possession of Lord Leighton for about half an hour, dear, if you don't mind. I've got something very serious to say to him. Dad, with the characteristic cowardice of his sex, has left it to me to say. It's—well, it's about a mummy: a female mummy, or, at least, I suppose I ought to say a mummy that was once a female—about five thousand years ago."

"My dear Niti——"

"No, no, don't interrupt me, for goodness' sake. It's too serious. It is really. We've had something like a tragedy here in the last few days, and things seem to have been, as you would say, a good deal mixed up ever since. I don't understand it a bit; but they have been."

"But, my dear Niti, what on earth can you have to say to Lord Leighton about a—a female mummy? What possible interest can a five-thousand-year-old corpse have for him?"

"Don't, Brenda, don't—at least not just now! Wait till I've told you, and then you'll see," said Nitocris, pressing her arm closer to her side.

"Lord Leighton is, as I think you know, an enthusiastic student of Egyptian antiquities. He was also, or thought he was, in love with my unworthy self. He found this mummy in a royal tomb at Memphis. He—well, I suppose, stole it—of course under the usual licence from the Khedive—and sent it home to Dad. Now comes the mystery. That was the mummy of Nitocris, the daughter of the great Rameses, and it was the dead image of my living self."

"Oh, but, Niti—what do you mean?"

"I don't know, Brenda. I wish I did. All I do know is that it was stolen that very night out of Dad's study in the Old Wing, and that I've got to tell Lord Leighton all about it. I'm sure Dad could have told him much better, only somehow he seems afraid."

"Oh, is that all—just the stealing of what was perhaps a very valuable relic? They try to steal much fresher corpses than that in the States if there are dollars in the business."

"Don't be brutal, Brenda! I know you don't mean it, and it isn't like you. Now, listen. Before he went to Egypt this time Lord Leighton asked me to marry him. I said 'No,' and for two reasons. I knew that he liked me very much—he always has done—and poor Dad took his liking for love and encouraged him: but I'm a woman and, I know, that liking isn't love—and then I love some one else. And now he, I mean Lord Leighton—loves some one else. Turn your face to the moon. Yes, you know who the some one else is. I'm so glad, for I do think you——"

"Niti, you're talking arrant nonsense for an educated young woman. I've only known His Lordship for a day, and how can you——"

"Because female Bachelors of Science and graduates of Vassar, whatever stupid people may say, have hearts as well as intellects, dear, and so they know. I seem to have had a kind of sixth sense given to me to-day, and, when you met Lord Leighton, I saw it, and I believe you *felt* it. I saw your eyes brighten and your face flush—only a little, but it did, and so did his. You know my belief in the Doctrine. You may have been lovers—perhaps wedded lovers—once upon a time, as they say in the fairy tales."

"How awful—no, I mean how wonderful—if it could only be true! And now, as you've told me all this, you might as well tell me who your some one else is."

"Really, Brenda, I thought you had more perception. He's there on the verandah smoking with your Lord Leighton."

"Oh! Then, of course, you're going to marry him?"

"I'm sorry to say Dad doesn't want me to. With all his genius and learning he is a perfect child in that sort of thing. He has no idea of Natural Selection. Now listen again, Brenda.. When I had to tell Mark that Dad wouldn't let me marry him, he picked me up out of a chair in the verandah there, where your father and mine are sitting, and kissed me three times."

"And I'll gamble ten cents that you kissed him back. That's Natural Selection, if I know anything about it. Niti, if that man—and he is a man—doesn't get killed in a fight, he'll marry you in spite of all the misguided scientific Dads on earth. Don't you worry. You've made me just happy. I'm not emotional that way, but I'd like to kiss you if the moon wasn't so bright. Suppose we go back and try to assist the kindly Fates a little bit?"

The Fates which, in some dimly-perceived fashion, seem to shape our little successive phases of existence, were certainly in a kindly mood that "lovely night in June." The two Professors had retired to Franklin Marmion's sanctum for the discussion of whisky and soda and the possibilities of physical manifestations of the Occult. Mrs van Huysman was frankly and comfortably sleeping in the deep, amply-cushioned armchair, and the two young men were almost as frankly pining for sweeter companionship than their own.

But the pairing off, which was so deftly managed by Nitocris, did not at first appear entirely satisfactory to them, yet a very few minutes' conversation sufficed to convince them of the wisdom of the arrangement. Brenda, with all the delicate tact which makes every highly-trained woman a skilled diplomatist, managed, not only to completely charm Merrill as a man who is in love with another woman likes to be charmed, but also to make him understand even more clearly than he had done how greatly the Fates had blessed him by giving him the love of such a girl as Nitocris; and then, by a few very deftly conveyed suggestions, she further gave him to understand that, so far as Lord Leighton had ever been an unconscious obstacle in his path, he was even now engaged in removing himself. Wherefore Commander Merrill enjoyed his smoke and stroll under the beeches a good deal more than he had anticipated.

More difficultly ambiguous, certainly, was the position in which Lord Leighton found himself with Nitocris, but here also her tact and perfect candour helped his own innate chivalry to accomplish all that was desirable with the slightest possible friction. She began by telling him, as she had told Brenda, of the mysterious stealing of the Mummy, and made a sort of apology for her father having deputed the telling of it to her—of course, in perfect innocence of the real reason for his doing so. He deplored with her the loss of what they both believed to be a priceless relic of the Golden Age of Egypt, but he passed it over lightly, chiefly for the reason that there was something in his mind just now that was much more serious than even the loss of the mummy of her long-dead namesake.

There had been a little silence between them after he had made his condolences, and then he said, with a hesitation which told quite plainly what was coming:

"Miss Marmion, I have a rather awkward confession to make to you—I have got to tell you, in fact, I think it is my duty to—well, honestly I really don't quite know how to put it properly, but—but—er, something has happened to me to-day that is a good deal more important to me, at least, than the disappearance of half a dozen royal mummies."

"Indeed?" said Nitocris, with a demurely perfect assumption of ignorance. "A good many things seem somehow to have happened to-day. It is something connected with that wonderful Adept's marvels, perhaps? They have certainly astonished most of us, I think."

"No," he replied, still a trifle hesitatingly, "it is nothing connected with him or his miracles, as far as I know, except that there was certainly something decidedly queer about the man and the impression he made upon one. Of course I have seen something like the same thing in Egypt and the Farther East; but he seemed quite what I might call uncanny. Still, that's not the point, although possibly it may have had something to do with it."

He hesitated again. She looked at him with a sideway glance, and said, almost in a whisper: "Yes?"

The moonlight was bright enough for him to see the notes of interrogation in her eyes, and he took the plunge.

"Miss Marmion, I once told you that I loved you and wanted you for my wife, and—and the real fact is that it—I mean I know now that it wasn't true—and so I thought I ought to tell you. You know, of course, that the Professor——"

"My dear Lord Leighton," she answered, with an air of quite superior wisdom, "my learned father is a very clever man in his own subjects: but I think I know a great deal more about this particular one than he does. You are quite right. You did not love me. You liked me very much, I have no doubt——"

"Yes, and so I do still, and always shall do, but——"

"But your liking was great enough to make you mistake it for love. Women's instincts are quicker and keener in these relations than men's are, and I saw that you did not love me as a real woman has to be loved, and, to be quite frank with you, some one else did. I like you very much, Lord Leighton, and I am going to go on liking you; but, you see, I could not give you what I had already given away. Now, you have told me so much that you ought to tell me a little more. How did your sudden enlightenment on that interesting subject come about?"

He was infinitely relieved by the absolutely frank and friendly way in which she had treated the whole subject, and so he had courage to reply with a laugh:

"In short, Miss Marmion, you ask me who the other girl is. Well, you certainly have a right to know, because, curiously enough, I might never have got to know her but for you— —"

"Is it Brenda?"

The question was whispered, and he replied in a whisper:

"Yes; do you think I have any chance?"

A cohort of wild cats would not have torn Brenda's secret out of her friend's soul, and so she replied in a tone that was almost judicious in its evenness:

"That, my friend, is a question that you can only get answered by asking another—and you must ask her, not me."

"Oh yes, of course I must," he said rather limply. "But she's so splendid— so beautiful, so exquisite—and—I do wish she wasn't so very rich. You see, even if I had the great good fortune to—to get her to marry me, I have lots for both; and, you know, the moment an Englishman with a title gets engaged to an American millionairess everybody says that he is simply dollar-hunting."

"That, unfortunately, is usually too well justified by the facts," she replied seriously. "But only the most idiotic and ignorant of gossips could possibly say that of you. Every one who is any one knows that the Kyneston coronet does not want re-gilding."

And then she went on, glancing sideways at him again:

"Still, as you know perfectly well, in matters of this kind, these very delicate diplomatic considerations, I do not care whether it is a question of fifty shillings a week or fifty thousand a year. You once paid me the very great compliment of offering me rank, position, and almost everything that a girl, from the merely material point of view could ask for. I refused, because I felt certain that you and I did not love each other—however much we may have liked and respected each other—as a man and woman ought to do, unless they become guilty of a great sin against each other. To put it in a very hackneyed way, we were not each other's affinities. I had already found mine—and I think, and hope, that you have found yours—and I wish you all the good fortune that you may, and, perhaps, can win."

"If is very, very good of you, Miss Marmion; but do you think you could—well, help me a little? I know I don't deserve it."

"No, sir, you do not," she laughed softly, because the other two were coming back on to the lawn. "I wonder that you have—I have half a mind to say the impudence—to ask such a thing. You have confessed your fickleness in an almost shameless way; and now you ask me to help you with the other girl! No, my lord: if I know anything of Brenda van Huysman's nature, there is no one who can help you except yourself. Of course she might——"

"Do you really think she might—I mean in that way?"

"Who am I that I should know the secrets of another woman's soul?" she replied, with unhesitating prevarication. "There she is. Go and ask her, and take my best wishes with you. Now I am going to talk to *my* affinity for a few minutes."

"So it was Merrill, after all!" he said to himself, as they joined the others. "Well, I'm glad. He's a splendid fellow; and she—of course, she's worth the love of the best man on earth—and I'm afraid that's not—anyhow, I'll have Miss Brenda's opinion on the subject before I go home to-night."

It now need hardly be added that the said opinion was not only entirely satisfactory, but also very sweetly expressed.

CHAPTER XIII
OVER THE TEA AND THE TOAST

The next morning there were, at least, three eventful breakfasts "partaken of," as it was once the fashion to say; one at "The Wilderness," one at the Savoy, and one at the Kyneston town house in Prince's Gate.

When Professor Marmion came down he was a little late, for he had done a long night's work, finishing his lecture-notes to his own satisfaction, or, at least, as nearly as he could get there. Like all good workers, he was never quite satisfied with what he did. When the maid had closed the door of the breakfast-room, he looked across the table at his daughter with a twinkle in his eyes, and said:

"Niti, before Lord Leighton left last night he had a talk with me, and you were partly the subject of it."

"And who might have been the other part of the subject, Dad?" she asked, with excellently simulated composure.

"That, Niti," he replied slowly, "I expect you know quite as well as I do. I am inclined to consider myself the victim of something very like a conspiracy."

"I think you are quite right, Dad," she replied, with perfect calmness. "But the chief conspirators were the Fates themselves. We others only did as we had to do. When you have solved that problem of N to the fourth, I think you will see that we could really have done nothing else, because, if you once crossed the border-line—the horizon which Professor Cayley spoke of, I mean—you ought to be on speaking terms with them."

Before he replied to this somewhat searching remark, the man who *had* crossed the horizon emptied his coffee cup, and set it down in the saucer with a perceptible rattle. Then he said more slowly than before:

"My dear Niti, there are other mysteries than N to the fourth. I only wish now to confess frankly to you that I have tried to solve one of them, perhaps the greatest of all, and ignominiously failed. I learnt a great deal last night from a young man to whom I thought I could have taught anything, and I got up this morning in a distinctly chastened frame of mind; and so, to make a long story short, if you like to drive into town and bring Commander Merrill back to lunch, I shall be very pleased to have a chat with him afterwards."

The next moment Nitocris was on the other side of the table, with her arm round her father's shoulders. She kissed him, and whispered:

"You dearest of dears! If I could have loved you any more, I would now, but I can't. I won't drive into town, because Brenda's coming out with Lord Leighton in her new motor to fetch me; at least, she will, if other papas have been as delightful as you have been."

He put his hand up and stroked her cheek with a gesture that was older than she was, and said with a smile which meant more than she could comprehend:

"Ah! so it *was* a conspiracy, after all! Well, dear, I hope that, for all your sakes, it will turn out a successful one."

About the same time Brenda was saying to her parents:

"Poppa and Mammy, I've got some news to tell you, and I've slept on it, so as to make quite sure about the telling."

"And what might that be, Brenda?" asked her mother, looking up a trifle anxiously. "Nothing very serious, I hope."

"Anything connected with the Marmions?" asked her father, in a voice that sounded as though it had come from somewhere far away. He had the *Times* propped up against the sugar basin on his left hand, and he had just read the announcement of Franklin Marmion's lecture for the following evening, and this was quite a serious matter for him.

"It's connected with them in this way," said Brenda, leaning her elbows on the table. "You and Uncle have wanted a coronet in the family, and you know that I've refused three, because the men who wore them weren't fit to respect, to say nothing about loving. Well, I've just discovered that I do love a man who has one coronet now, and will have another some day, unless something unexpected happens to him; but mind, it's the man I love and want to marry, and I'd want to do it just the same if he was still the same man he is, and hadn't either a coronet or a dollar to his name."

"That's like you, Brenda, and it sounds good," said her father, tearing his attention away from the alluring title of Franklin Marmion's lecture. "Now, who is it?"

"If it was only that nice young man, Lord Leighton!" said Mrs van Huysman, in a voice that sounded like an appeal against the final judgment of human fate, "but, of course, he's— —"

"No, Mammy, that's just what he's *not* going to do," exclaimed Brenda, sitting up and clasping her hands behind her neck. "Nitocris Marmion is in love with some one else, and Lord Leighton is in love with me—at least he said so last night at 'The Wilderness,' and I don't suppose he'd have said it if he hadn't meant it—and I told him to go and ask his Papa: and now I'm going to ask my Poppa and Mammy if I may be Lady Leighton soon, and, perhaps, some day Countess of Kyneston. You see, Lord Leighton is just a viscount now— —"

"What, just a viscount!" exclaimed Mrs van Huysman, getting up from her chair and putting a plump arm round her neck. "Just a viscount—and heir to one of the oldest peerages in England! Oh, Brenda, is it really true?"

"I guess Brenda wouldn't say it if it wasn't, and that's about all there is to it," said her father, putting his long arm out over the table. "I congratulate you, my girl. Mammy and I may have been a bit troubled over some of those other refusals of yours, but you seem to have known best, after all: and I reckon your Uncle Ephraim'll think the same. Lord Leighton's a man right through. He wouldn't have done what he has done if he hadn't been. Shake, child, and— —"

Brenda "shook," and then, without another word, she got up and hurried out of the room.

"The girl's right!" said Professor van Huysman, as the door closed behind her; "and if I'm not a fool entirely, she's found the right man."

"Hoskins, you can leave that to a well-brought-up girl like Brenda all the time. She *is* right, and all we've got to hope for now is that the Earl will be right too," said his wife somewhat anxiously.

"He's just got to see our girl and then he will be, unless he's a natural born idiot, which, of course, he couldn't be," replied Brenda's father in a tone of absolute conviction. "Now, I wonder what that man Marmion's going to let loose on us to-morrow night?"

"Good morning, sir," said Lord Leighton, as his father came into the breakfast-room at about the same time that Brenda left the other room in the Savoy.

"Good morning, Lester," replied the Earl of Kyneston, as father and son shook hands in the old courtly fashion which, within the last half century, has gone out of vogue save among those who have ancestors whose record is a credit to their descendants. "You are looking very well and fit—and there is something else. What is it? Had you a very pleasant evening yesterday at 'The Wilderness'? Has Miss Marmion revoked her decision after all?"

"No, sir," said his son, looking at him with brightening eyes; "but she convinced me that I had thought myself in love with the wrong girl—and the other girl was on the lawn at the same time, talking with the man that Miss Marmion was, and is in love with, and will be always, I think."

"And the other young lady, Lester—because, of course, she is a lady, I mean in our sense of the word, much misunderstood as it is in these days?"

"She is Brenda van Huysman, sir."

"Oh, the Professor's daughter.—I mean the other Professor's daughter. A very good family. Her father is a distinguished man, and, if I remember rightly, a Van Huysman was one of the first colonisers of New England about four hundred years ago. It is the same family, I suppose?"

"Yes, sir; I can vouch for that."

Nitocris had given him the whole history of the family, and so he was sure of his facts.

"Lester, I congratulate you," replied his father, taking his arm, as they were accustomed to. "While you have been away digging among those Egyptian tombs and temples, this girl has refused at least three coronets, and one had strawberry leaves on it; so she loves you for yourself. That is good, other things being equal, as I think they will be in this case. Now, we will go to breakfast, and you shall tell me the whole story. I have not heard a real love story for a good many years."

CHAPTER XIV
"SUPPOSED IMPOSSIBILITIES"

It was only to be expected that the announcement of a lecture with such an alluring title by such a distinguished scholar and scientist as Professor Franklin Marmion should fill the theatre of the Royal Society, as the reporters said tritely but truly, "to its utmost capacity."

The mere words, "An Examination of Some Supposed Mathematical Impossibilities," were just so many bomb-shells tossed into the middle of the scientific arena. The circle-squarers, the triangle-trisectors, the cube-doublers, the flat-worlders, and all the other would-be workers of miracles plainly impossible in a world of three dimensions jumped — not incorrectly — to the conclusion that their favourite impossibility would be selected for examination, and, perhaps — blissful thought! — demonstration by one of the foremost thinkers of the day, to the lasting confusion of the scoffers. Learned pundits of the old school, who were firmly convinced that Mathematics had long ago said their last word, and that to talk about "supposed impossibilities" was blasphemy of the rankest sort, came with note-books and a grim determination to explode Franklin Marmion's heresies for good and all. Dreamers of Fourth Dimensional dreams came hoping against hope, for the Professor was known to be something of a dreamer himself; and added to all these there assembled a distinguished company of ladies and gentlemen who looked upon the lecture as a "function" which their social positions made it necessary for them to patronise. The reader's personal friends and acquaintances, including Prince Oscarovitch and Phadrig, were naturally among the most anxiously interested of the Professor's audience.

It is almost needless to say that Hoskins van Huysman had donned all his panoply of scientific war, and had armed himself with what he believed his keenest weapons; and that Professor Hartley looked with amused confidence to a veritable battle royal of wits when the lecture was over and the discussion began. The Prince and Phadrig were keenly anticipative, and the latter not a little nervous.

A verbatim report of that famous lecture would, of course, be out of place in these pages. If Professor Marmion's words of wonder are not already written in the archives of the Royal Society, no doubt they will be in the fullness of time when the minds of men shall have become prepared to receive them. Here we are mainly concerned with the results which they produced upon his audience. Certain portions may, however, be properly reproduced here.

When the decorous murmur of applause which greeted the President's closing sentences had died away, and Franklin Marmion went to the reading-desk and unfolded his notes, there was a tense silence of anticipation, and hundreds of pairs of eyes, which had some of the keenest brains in Europe behind them, were converged upon his spare, erect figure and his refined, clear-cut, somewhat sternly-moulded face.

"Mr President, my lords, ladies, and gentlemen," he began, in his quiet, but far-reaching tones. "The somewhat peculiar title which I have chosen for my lecture was not, I hope I need scarcely say, selected with a view of arousing any but that intelligent curiosity which is always characteristic of such a distinguished audience as that which I have the honour of addressing to-night. I chose it after somewhat anxious consideration, because I am aware that the bulk of opinion in the world of science strongly insists upon the finality of the axioms of mathematics, and therefore it was with no little hesitancy that I approached such a subject as this. I am well aware that, in the estimation of most of my learned *confrères* and fellow-seekers after scientific truth, to suggest those axioms may not embody final and universal truth is, if I may put it so, to lay sacrilegious hands on the Ark of the Scientific Covenant."

A low murmur, prelude of the coming storm, ran through the theatre, and Professor van Huysman permitted himself to snort distinctively, for which he was very promptly, though quietly, called to order by his daughter, who was sitting in front of the platform between him and Lord Leighton. Franklin Marmion paused for a moment and smiled ever so faintly. Nitocris looked round at the now eager audience a trifle anxiously, for she had a fairly clear idea of the trouble that might possibly be ahead. Her father went on as quietly as before:

"Of course, every one here is aware that the great Napoleon once said that the word 'impossible' was not French. I need not remind such an audience as this that more than one distinguished student and investigator has suggested that it also may not be scientific."

The murmur broke out again, and Hoskins van Huysman blew his nose somewhat aggressively. His scientific bile was beginning to rise. He disapproved very strongly of the tone which his rival had begun. Its quiet confidence was somewhat ominous. The lecturer continued without this time noticing the interruption, and proceeded to give a lengthy and learned but singularly lucid *resumé* of the more recent progress in the higher mathematics and the deeply interesting speculations to which it had given rise. This, with certain demonstrations which he made on the great black-board beside him, occupied nearly an hour. When he had finished there was another murmur, which this time was wholly of applause, for this part of the lecture had not only been masterly but entirely orthodox. Then silence fell again, the silence of expectant waiting, for every one felt that the "Examination" was coming now. He began again in a slightly altered voice.

"What I have just been saying was necessary to my subject as far as it went, but for all that it was chiefly introductory to what I am now going to bring to your notice. But this is a matter rather for illustration and discussion than for mere disquisition. Therefore, to save your time as much as possible, I will proceed at once to the illustration, and then we will have the discussion."

Professor van Huysman snorted again, even as a war-horse that snuffs the fray. This time Franklin Marmion seemed to recognise the implied challenge, for he looked round the crowded theatre with a curious smile, which seemed to say: "Yes, gentlemen, I see that some of you are getting ready for a tussle. I am in hopes of being able to oblige you."

"Now," he continued, "it is generally conceded that an ounce of practice is worth a good many pounds of precept, so I will get to the practice. I need hardly remind you that ever since mathematics became an exact science, three problems have been recognised as impossible of solution—trisecting the triangle, squaring the circle, and doubling the cube. I have now the pleasure of announcing that I have had the great good fortune to discover certain formulæ which, so far, at least, as I can see, make the solution of those problems not only possible, but comparatively easy—to those who know how to use them."

As he said this, Franklin Marmion looked directly at Hoskins van Huysman. He was the challenger now, and there was a glint in his eyes and a smile on his lips which showed that he meant business. The American writhed, and had it not been for Brenda's gently but firmly restraining hand, he might have jumped to his feet and precipitated matters in a somewhat embarrassing fashion. The chairman looked up at the lecturer with elevated eyelids which had a note of interrogation under each of them, and then there came that sound of shifting in seats and breathing in many low keys which denotes that an audience has been wound up to a very tense pitch of expectation. If a smaller man had said such words to such hearers some one would have laughed, and then would have burst forth a storm of derision. But the keenest critic had never found Franklin Marmion wrong yet, and he had far too great a reputation to permit himself to say in such a place that which he did not seriously mean. So the hum died down as he went to the black-board, and Nitocris looked at Merrill with something like fear in her eyes.

"If he does that," whispered Phadrig to the Prince in Russian, "the story that Pent-Ah and Neb-Anat told will be true—which the High Gods forbid!"

"As the trisection of the triangle is, perhaps, the simplest of the three problems," said the lecturer, with almost judicial calmness, "we will, if you please, begin with that. I hope that gentlemen who have brought note-books with them will be kind enough to follow my calculations and check any error that I may make."

But a good threescore note-books, pencils, and stylographic pens were out already, and hundreds of eyes were eagerly fastening their gaze on the black-board, their owners desperately anxious to detect the first slip in the demonstration. The demonstrator drew an isosceles triangle rapidly, and without speaking filled the remainder of the board with formulæ. The almost breathless silence was broken only by the click of the chalk on the board and the scratching of pencils and pens on paper. When he had finished he ran through the calculations aloud, and said in the most commonplace voice:

"Now, gentlemen, if, as I hope, you have found my working correct, I may draw the two lines which will trisect the triangle."

He drew them, and then, as calmly as though he had done nothing more than cross the much-trodden *pons asinorum*, he told two attendants to take the board down and put it in front of the platform; then, while they were lifting another on to the easel, he said:

"As those who have followed me would no doubt like a little time to revise the figures, I will go on with the next problem, which will be our old friend, or enemy, the squaring of the circle."

The second board was filled with diagrams and formulæ as rapidly as the first.

"There is the demonstration, gentlemen," he said, as the attendants placed it beside the other in full view of everybody. "Now, as time is shortening, I will get on with the third problem."

The chalk began to click again, and the pens and pencils scratched on to the accompaniment of murmurs and whispers and occasional grunts and snorts of incredulity. By a master-stroke of strategy Franklin Marmion had, in placing the three demonstrations of the long-supposed impossible before them in quick succession, kept the learned, but now utterly bewildered mathematicians so busy that they literally had not time to begin "the trouble" which Brenda was now actually dreading. Her father's face, bent down over his note-book, was getting more terrible to look upon every moment. The mere fact that he had not uttered a sound since the demonstrations had begun was sufficiently ominous, for it meant that he was puzzled — perhaps even beaten — and if that was so, she dreaded to even imagine what might happen. On the other hand, Nitocris felt her spirits rising as she looked round and saw the many learned heads bending and shaking over the note-books, each owner of them working at high pressure to win the honour of first finding the error which all firmly believed must exist, and which none of them could detect.

When he had finished his third demonstration, Franklin Marmion, without interrupting the hard thinking that was going on, took a chair by the side of the President, poured out a glass of water, and waited for results.

"Marmion, what is this white magic that you have been springing upon us?" whispered the presiding genius of the learned assembly, looking up from several sheets of paper which he had been rapidly covering with formulæ. "These things are impossible, you know — unless, of course, you have got a good deal farther than any of us. And yet the calculations are correct as far as I can follow them, and no one else seems to have hit on any error yet. I must confess, though, that these progressives of yours are too deep for me. I can follow them, and yet I can't. At a certain point they seem to elude me, and yet the calculations are rigidly right. It's almost enough to make one think you had done what Cayley once told us in this room some one might do some day."

"My Lord," replied Franklin Marmion, almost inaudibly, "I began my address by remarking, as you will remember, that perhaps, after all, the word 'impossible' might not be scientific."

Their eyes met, and the President, than whose there was no greater name in the higher realm of learning, saw something in Marmion's which sent a little chill through him, and that something told him that he was in the presence of a superior being.

"Dear me!" he murmured, looking down at his papers again, "the age of miracles is not past, after all — in fact, it is only just beginning."

"It is re-beginning, my Lord—for us," came the reply, in a voice which seemed to come from very far away.

The President did not reply. As a matter of fact, he had no reply ready, and he had something else to do. He rose, and said in a somewhat constrained voice:

"Ladies and gentlemen, Professor Marmion has shown us some very strange demonstrations which have certainly amply justified the title which he selected. A good many gentlemen, and some ladies as well, I am glad to see, have followed his calculations very carefully. I have done the same myself, but I am bound to confess that I have not been able to find any error. I think I shall be right in saying that no one will be more pleased than the learned and—er—gifted lecturer to hear that some one else has been able to do so."

Franklin Marmion bowed his assent, and a faint smile flickered across his clean-shaven lips. The next instant Professor van Huysman was on his legs, note-book in one hand and stylo in the other. All the fresh colour had gone out of his face; his eyes were burning, and his lips were twitching with uncontrollable excitement.

"My Lord," he began, in a voice that even Brenda hardly recognised, "like yourself, I have been unable to find any actual error in the lecturer's demonstrations of which I will take permission to call the possibility of the impossible; in other words, that a contradiction in terms can be true and false at one and the same time. That, my Lord, and ladies, and gentlemen," he went on, raising his voice almost to a shout, "is still, and, I hope, in the interests of true science, and not adroit jugglery with figures and formulæ, will ever remain, another impossibility. Professor Marmion has apparently trisected the triangle, squared the circle, and doubled the cube. It may be that he has persuaded some present that he really has done so; but, again, in the interests of science, I desire to protest against the way in which these demonstrations have been sprung upon us. Calculations which he has doubtless taken months to elaborate, he has asked us to test in a few minutes. For myself, I decline to accept them as true, and I hope that others will do the same until we have had time to satisfy ourselves that the hitherto impossible has been made possible."

He sat down, breathing hard and white with anger and excitement, and then the trouble began. The trisectors, the circle-squarers, and the cube-doublers, had seen their long-flouted theories proved to demonstration by one of the most learned and responsible men of science in the world, and one of their most sarcastic and hitherto successful flouters had been compelled to confess that he could find no flaw in the calculations of this mathematical Daniel so unexpectedly come to judgment. They did not understand his proofs, but that was no reason why they should reject them, and so they rose as one man in support of their champion to demand that Professor van Huysman should withdraw his imputations of jugglery. He sat still, and shook his head. He was too disgusted and bewildered to do or say anything more until he had made a searching analysis of these diabolical formulæ.

But there were others who wanted to have their say in defence of scientific orthodoxy, and they had it—and the rest was a chaos of intellectual conflict until, at the end of nearly an hour, the President, who now saw with clearer eyes than any of the disputants, rose and put an end to the discussion by remarking that they had not the whole night before them, and that all that Professor Marmion had said and done would be published in the scientific papers; further, that such a controversy would perhaps be more profitably conducted in print than by word of mouth. Such a course would give every one ample leisure to work out the problems in the light of the new demonstrations, and also give a much better prospect of reaching a logical, and therefore just, conclusion than a discussion in which haste, and possibly pre-conceived opinions, from the influence of which no human being was really free, could possibly promise.

This, of course, put an end to the matter for the time being, and, after the usual votes of thanks and acknowledgments, the distinguished company dispersed—amused, mystified, gratified, bewildered, and exasperated: but, saving only four of its members, with no idea of the effect which that evening's proceedings were destined to have upon the fate of Europe, perhaps of the whole human race.

CHAPTER XV
THE ADVANCEMENT OF NITOCRIS—
THE RESOLVE OF OSCAROVITCH

Franklin Marmion and Hoskins van Huysman parted that evening in what may be described as a state of armed neutrality, but with more cordiality than Brenda, at any rate, had hoped for. Still, they were both gentlemen, and, moreover, the American scientist was honestly looking forward to the discovery of some fatal flaw in the reasoning of his English rival which should leave the final triumph with him—and such a triumph would be not only final but crushing.

Brenda whirled her father and Lord Leighton—who, of course, sat beside her in front as she drove—off to supper; Merrill went to his club to ruminate happily for an hour; and the hero of the evening and his daughter drove home almost in silence, and it was a silence for which there was a very sufficient reason. Such people do not talk about trivialities when they are thinking about much more serious concerns.

After supper Nitocris followed her father into the study, as he quite expected her to do, and when she had shut the door, she faced him and said in a voice that was not quite her own:

"Dad, there seems to me to be only one explanation of what you did to-night. I know enough mathematics to see that it is the only one. If you tell me that I am wrong, of course I shall believe you—and then I shall ask you how else you did it."

As she spoke he felt that his soul was asking itself a momentous question. She had guessed—or did she already know?—the Great Secret. And, if either, was she herself near enough to the dividing line between the two worlds for him to tell her the truth?

He sat down in the chair before his writing-table and stared hard at his plotting-pad for a few moments. Then he looked up at her and saw the answer.

"Niti," he said slowly, and with a little halt between the words, "you have asked me a question which I think some one else must answer, if it can be answered at all. Look behind you!"

She turned swiftly, and there, almost beside her, stood—not the Mummy, but the Queen, her living other-self, royal-robed and crowned as she had been in the dim past, which was now again the present.

Would she flinch or faint, or cry out with fear? If her unconscious feet had not advanced very near to the Border she would certainly do one or the other. Indeed, it was with an inward quaking of fear for her that her father had told her to turn. It might well have meant the difference between sanity and insanity, knowing what she already did of the Mummy and its mysterious disappearance. But no: there before his eyes was worked again the miracle which had already been worked in his own case, though now it was, if possible, even more marvellous than it had been before. As Nitocris turned she uttered a low cry of wonder and recognition, and held out both hands to her other twin-self. The Queen took them, and said in the Ancient Tongue, which now she understood again after many centuries:

"Welcome, thou who wast once myself, into this larger life to which the Perfect Knowledge hath led thee: where Time is not, and that which was, and is, and shall be are the same! Thou hast yet many days, as men call them, to live in that limited life known as mortal, and so the mortal lot, with its perils and sorrows and joys, shall yet be thine: yet, although, if the High Gods will it so, that life shall end and begin and end again many times, thou hast already won through the shadows which bound that little life into the light of the Day which knows not dawn nor noon nor night. I who was, and thou who art, are one again!"

Then came silence. Franklin Marmion saw the two kindred shapes merge into each other. He closed his eyes for a moment, as he thought, and when he opened them again he was alone. He looked at the clock, and saw that it was after four.

"Dear me!" he said, getting up with a shake of his shoulders, "I must have fallen asleep. Where's Niti? Why, of course, she has been in bed for hours, and it's about time that I got there, too."

When they met before breakfast Nitocris said to him:

"I had a very strange experience last night, Dad. I either saw, or dreamt I saw, the Mummy alive again, robed and crowned like a queen of ancient Egypt; and then we seemed to become the same person, and I remembered that I had been Queen Nitocris of Egypt once. Then I found myself alone—so very much alone—in a new world which was still like this one, only there wasn't any time. I had another sense which made me able to see past, present, and future all at once, and here and there, and up and down, and something else were all the same, and yet it did not seem in the slightest strange to me, so I suppose it was a dream."

"It was no dream, Niti," said her father, looking at her with grave eyes. "Last night, as we have to say in the state of Three Dimensions, you had your first glimpse of the state of Four. I saw what you did."

"Ah!" she replied, without any sign of astonishment. "Then that is why I was able to understand your demonstrations last night when all the rest were puzzled. I didn't think I quite did then, however, but I see now that I did. And so I and Her Majesty are really one and the same! It ought to seem very wonderful, but somehow it doesn't in the slightest."

"I don't think that anything will seem wonderful to you now, Niti," was the quiet response. "But as we are at present on the lower plane of existence, it will be necessary for us to go to breakfast."

Oscarovitch and Phadrig went back after the lecture to the Prince's flat in Royal Court Mansions, which, as a bachelor and a bird of passage, he found much more convenient in many ways than a house. He ordered his Russian servant to make coffee for his guest, and mixed a stiff brandy-and-soda for himself. He wanted it, for the experiences of the evening had shaken even his nerves not a little. He was essentially a man of power, both physically and mentally, of boundless ambitions and iron will, vast knowledge of the world, as he knew it, and of very high intellectual attainments; but the cast of his mind was absolutely material, and therefore he both hated and feared anything which appeared to transcend the material plane to which his mental vision was at present entirely confined.

When the servant had left the room after bringing the coffee, he gave Phadrig a cigar, lit one himself, and said through the first puffs of smoke:

"Phadrig, you know, or pretend to know, more about these things than I do, or want to do: but, still, just now I want you to tell me honestly if you believe that Professor Marmion did really solve those problems to-night. I ask you because I admit that the solutions went beyond the range of my mathematics."

"Highness," replied the Egyptian, speaking slowly and almost reverently, "he did. There is not, I think, another man on earth now who could have done so; but for those who had eyes to see there could be no doubt, and you will find that, though he has many rivals and will have countless critics, not one will be able either to explain his solutions or find a flaw in them."

"You did a few things that I should not have thought possible the other day, which you claimed to be really miracles. Now, if they were, I suppose you can explain Professor Marmion's?"

"There are no miracles, Highness: only the results of higher knowledge than that which they who see them possess. That is why what I did seemed like miracles to those who watched. But this Franklin Marmion, as he is called in this life, has attained to a higher knowledge than mine, wherefore I am able only to understand imperfectly, but not myself to do, that which he does. Yet, as the High Gods live, he did this thing; and to do it he must have passed to the higher life through the gate of the Perfect Knowledge."

"In other words," said the Prince, after a big gulp of his brandy-and-soda, "that he has solved that infernal problem of the fourth dimension you have had so much to say about. Now, granted that he has done so, what does it amount to as regards our world—the world of practical thought and real action, I mean?"

"All thought is practical, Highness," replied Phadrig, "since there can be no action which is intelligent without thought. Wherefore, the higher the thought the more potent the action, and so he who has the Perfect Knowledge has also the Perfect Power."

"Then, do you mean to tell me seriously—and I can hardly think that you would trifle with me—that this man is now practically omnipotent, as far as we lower beings, as you seem to call us, are concerned?"

"Only the High Gods are omnipotent, Your Excellency; but, if I have seen rightly, he is as a god to us of the lower life, and therefore I would pray you again to utterly relinquish your lately and, as I have dared for your sake to say, rashly-formed designs to make the Queen who was, and his daughter that is, the sharer of your future throne. Is not the Princess Hermia noble and fair enough?"

"No, by all your gods, no!" exclaimed the Prince passionately. "Since I have seen the woman who, as you say, was once Queen of Egypt, there is, and shall be, no other consort for me. And who are you to advise me thus? Are you still the same man who made the condition that, if you used your arts, whatever they may be, to place her in my power, she should be, not only my Empress, but also Queen of Egypt? What has changed you? What has made you faithless to the promise that you gave me in exchange for mine? If you have forgotten that, do not also forget that we Russians have a short way with traitors."

"What has changed me, Highness," replied Phadrig, ignoring the threat, "is the knowledge that I have gained to-night. Though you believe me or not, the debt which I owe you makes it my duty to warn you. The matter stands thus: Nitocris, the daughter of Franklin Marmion, was the Queen. For all I know, she also may have attained to the higher life, and is therefore the Queen still, though that is a mystery beyond my comprehension; but I do know now that her father has attained to it, and that for this reason, unless you put this new-found love out of your heart, you will bring yourself within the sphere of this man's power—a power mighty enough to wreck every scheme you have ever shaped, and to doom you to a fate more horrible than mortal brain could conceive. You would be as a man who strove against a god."

"You may believe what you are saying, Phadrig, and I dare say you do," exclaimed the Prince again. "I don't, because I can't; but even if I did, I would claim your promise. I love this Nitocris, Queen or woman, and neither man nor god shall keep her from me, willing or unwilling. As for the Princess Hermia—well, her husband is not dead yet."

"Better he dead and his widow your wife, as was planned, Highness, than that you should dare the power of one who has attained to the Perfect Knowledge," said the Egyptian, with all the earnestness of absolute conviction. "But my duty is done. I have warned you of that which you cannot see for yourself. I have done it to my own sorrow and the destroying of my own dream; but my promise is given, and I will keep it, even to a fate that may be worse than death."

The Prince drained his glass and laughed.

"Well said, my ages-old adept, as you think you are! You shall follow me, for I will go on now even to death, or what there may be worse behind it, if I can only take my beautiful Queen with me. Yes, I swear I will, by God—if there is one!"

So by his ignorant blasphemy Oscar Oscarovitch, who once was Lord of War in Egypt, for the love of the same woman, fixed his fate for this life, and for many that were to come after it.

CHAPTER XVI
THE MYSTERY OF PRINCE ZASTROW

Events now began to move with an almost bewildering rapidity, at least, so far as they affected the immediate temporal concerns of Nitocris and her father. For days and weeks a furious storm raged round the famous lecture, and the atmosphere of the scientific world was thick with figures and formulæ, diagrams and disquisitions; but since none of the learned disputators proved himself capable of detecting the slightest flaw in the lecturer's mathematics, it had very little interest for him, and therefore has none for us. In fact, so little did he seem concerned with the tempest he had raised, that a few days later, to the astonishment and chagrin of his baffled critics, he and Nitocris bade adieu to their more intimate friends and disappeared on a wandering trip of undetermined destination for change of air and scene and a much-needed holiday for the over-worked Professor. At least, that is the reason which Nitocris gave to Lord Leighton and the Van Huysmans, and the few others to whom she thought it necessary to give any explanation at all.

The day before they left, Merrill lunched at "The Wilderness," took a fitting leave of his lady-love and his prospective father-in-law, and departed to join his ship, slightly mystified, perhaps, by recent happenings, but still believing himself with sufficient reason to be the happiest and most fortunate Lieutenant-Commander in the British Navy.

The true reasons for the sudden departure of the now more than ever famous Professor and his beautiful daughter from the scene of his latest and most marvellous triumph may be set forth as follows:

On the evening of the third day after the lecture Franklin Marmion was going back by train to Wimbledon after a long day at the British Museum among the relics of Egyptian antiquity—which, as may well be understood, he studied now with an interest of which no other man living could have been capable; and as soon as he was seated in a comfortable corner, and had his pipe going, he opened his *Pall Mall Gazette*, and, as was his wont on such occasions, began with the leading article and read straight along through the Special Article and the Occ. Notes, until he came to the news of the day, skipping only the financial news and quotations, which, under his present changed conditions of existence, he dare not trust himself to read lest he might be tempted by the unrighteousness of Mammon, a form of idolatry which he had always heartily despised.

The first item on the news page was headed in bold type:

"MYSTERIOUS DISAPPEARANCE OF A RULING GERMAN PRINCE.

"SUSPICION OF FOUL PLAY.

"IMPORTANT STATE PAPERS VANISH WITH HIM.—SPECIAL.

"In spite of the most rigorous censorship of the Press Bureau, it has now become a matter of practical certainty that Prince Emil Rudolf von Zastrow, the youthful and very capable ruler of Boravia, who, during the last two or three years, has become one of the most brilliant figures in European society, has disappeared under circumstances so strangely mysterious as to suggest some analogy with the tragedy of which the unhappy Prince Alexander of Bulgaria was the central figure.

"The facts, so far as they have been ascertained, are briefly as follows:—Up to about a fortnight ago, the Prince was living in semi-retirement with his consort, the Princess Hermia, in his picturesque Castle of Trelitz, which, as every one knows, looks down over the waters of the Baltic from a solitary eminence of rock which rises out of the vast forests that cover the rolling plains for leagues on the landward sides. It will be remembered that every year since his accession, the Prince has been wont to retire to this famous hunting-ground of his to enjoy at once the pleasures of the chase and the society of his beautiful young consort in peace and solitude after the whirl of the European winter season. As far as is known, the only guests at the Castle were the Count Ulik von Kessner, High Chamberlain of Boravia, who is believed to have been present on business of State, and Captain Alexis Vollmar, of the 55th Caucasus Regiment, at present attached to the Imperial Headquarter Staff at St Petersburg. Captain

Vollmar, in addition to being a brilliant young officer, is also a scion of two of the wealthiest and most aristocratic families in Russia.

"It is now fully established that on the evening of the 6th of this month—that is to say, nearly three weeks ago—the Prince and his two guests returned after a long day in the forest, and that the Prince retired to rest very shortly before supper. From that day to this he has never been seen, either at home or in society. What makes the disappearance more strangely striking is the fact that the Prince, who is Colonel of the 28th Pommeranian Regiment, did not put in an appearance at the recent review in the Kaiserhof when the German Emperor held his usual inspection. Although it was obvious that His Majesty was both puzzled and annoyed by his absence, no official explanation of it has been given, and all information on the subject is rigidly withheld. Our own comes from a personal friend, and, as far as it goes, may be absolutely relied upon."

For some reason or other, which, after his recent experiences, he thought it would be as well not to try and fathom for the present, these few paragraphs made a strangely persistent impression on him. When he got home he gave his evening papers as usual to his daughter, and at dinner the Zastrow mystery was the chief, in fact almost the only, topic of conversation.

"Yes, it certainly is very extraordinary," said Nitocris. "The papers make mysteries enough out of the disappearance, of the most everyday, insignificant persons, who were probably only running away from their debts or their domestic troubles, but for a real Prince to utterly vanish like this—that certainly looks like a little more than an ordinary mystery. And I suppose," she went on, after a little interval of silence, "if there really has been foul play—I mean, granted that Prince Charming, as all the Society papers got to call him, has been spirited away for some hidden reason of State or politics and is never intended to see the light of day again, who knows how many secrets may be connected with this affair which might be like matches in a powder magazine? And—Oh yes—why, Dad, it was this same Prince Zastrow who has been mentioned by most of the best European papers as the only possible Elective Tsar of Russia if the Romanoffs are driven out by the Revolution, and the people go back to the old Constitution. In fact, some of them went so far as to say that nothing but his selection could prevent a scramble for the fragments of Russia which could only end in general conflagration."

"Yes, of course I do," replied her father. "But what an atrocious shame, if it is so! One of the most popular of the minor princes of Europe spirited away, and perhaps either murdered or thrown into some prison or fortress, where he will drag out his days and nights in solitude until he goes mad: a young, bright, promising life ruined, just because he happens to stand in the way of some unscrupulous ambition, or vile political intrigue!

"It would be a crime of the very first magnitude, that is to say, of the most villainous description, and all the more horrible because it would be committed by people in the highest of places. Really, Niti, it is enough to make one think that there ought to be some higher power in the world capable of making these political crimes impossible. The inner history of European politics—I mean, the history that doesn't get into books or newspapers—would, I am certain, prove that quite half the wars of the world, at least during the period of what we are pleased to call civilisation, would have been avoided if some means could have been found of putting an end to the miserable personal ambitions and jealousies which have never anything to do with the welfare of nations, but quite the reverse. I shouldn't wonder if poor Prince Zastrow has been the victim of something of the sort. It is quite possible that expiring Tsardom had a finger in the pie. At any rate, there was a Russian officer in the Castle the day he disappeared. I should very much like to see the sort of explanation *he* could give of the affair, if he chose."

"But is there not such a power in the world now, Dad?" asked Nitocris, looking across the table at him with a peculiar smile.

He looked back in silence for a moment or two. Then he replied slowly:

"I see what you mean, Niti. Of course, I suppose we shall be able to read each other's thoughts now, or even converse without speaking, or when we are out of earshot of each other. The same idea came to me while I was reading the account of this affair in the train; but should I, or, rather we, be doing right in interfering actively in the transactions, political and otherwise, of the world—by which I mean, of course, the state of three dimensions? It would be a terrific responsibility. Remember what tremendous powers we are capable of wielding by simply—it is so very simple now—simply transferring our personalities to the higher plane. What if we were to do wrong? We might involve the whole world in some unspeakable catastrophe."

"And which do *you* consider to be the greatest catastrophe, or, perhaps I ought rather to say the greatest evil, that has ever afflicted the world, Dad?" she asked, with just a suspicion of a smile in her eyes, though her lips were perfectly serious.

"Oh, war, of course!" he replied, with his usual emphasis when he got on to that topic. "What was I saying only just now about personal intrigues and ambitions that make war? What have I always thought about war? It is the most appalling curse——"

"Then, Dad," she interrupted in her sweetest tones, "do you think that, supposing we possess these wonderful powers, they could be better used than in preventing any war which may possibly arise out of this disappearance of Prince Zastrow, and so convincing those who are wicked enough to plunge the human race into blood and misery that henceforth all wars of aggression and ambition will be impossible?"

"Yes, you are right as usual, Niti," he exclaimed, getting up. "Now you go and think about it all, and give me your advice in the morning. I want to get away now and work out an intelligible solution of those three problems—if I can make it so—for the benefit of Van Huysman and the rest of my respected critics. When I've done that, we'll be off to the Continent or somewhere——"

"And see what we can make of the Zastrow Mystery, perhaps!" said Nitocris. "Good-night, Dad. I want to do some thinking, too."

He went to his study and set to work upon a development of the demonstrations with which he had astounded not only London, but the whole civilised world.

But it was no good to-night. The ideas would not come. Over and over again he picked up the threads of his arguments, only to drop them again. At last, in something like wondering despair, he muttered:

"Confound the thing! I almost had it last night, and now I seem as far away from it as ever. What on earth can be the matter with me?"

He put his elbows on the table, took his head between his hands, and stared down at the pages covered with angles and circles, chords and curves, and wildernesses of symbols, which were scattered about his desk. As he stared at them they seemed somehow to come together, and the lines and curves arranged themselves in symmetrical shapes, until they developed from diagrams into pictures; and as they did so he found himself forgetting all about the problems, and thinking only of the strange vision which seemed to be unfolding itself among the scattered papers before him. The straight lines became the walls and turrets of one of those two-or three-hundred-year-old German country houses, half castle, half mansion, which every explorer of the bye-paths of the Fatherland has seen and admired so often. The curves became long, sweeping stretches of sandy bays, fringed with other curves of breaking rollers; and as the picture grew more distinct, one great circle embraced a whole perfect picture of land and seascape—land dusky and forest-covered in the southward half; and the misty sea, island-dotted, wind-whipped, and foam-flecked, to the northward.

The castle stood on the top of a somewhat steeply sloping hill about five hundred feet above the sandy shore, on which the breakers were curling a couple of miles away. The hill was covered with thick-growing firs from the plain to the castle wall, but two broad avenues ran in straight lines, one to seaward, and the other down into the depths of the vast forest, until it opened on to the post road, which afforded the only practicable carriage route to the station of Trelitz on the main Berlin-Königsberg Railway.

The longer he looked, the more surprisingly distinct the picture became, and, curiously enough, the less his wonder grew. He saw three men on horseback riding at a canter up the avenue from the forest. Their costumes showed plainly enough that they had just come back from the chase. As they rode on they seemed to come quite close to him, until he could see their features with perfect distinctness. By the changing expression of their faces he could tell they were laughing and chatting; but, singularly enough, he could not hear a word that they were saying, which, considering the minuteness with which he saw everything, struck him as being distinctly curious.

He watched them ride up to the old Gothic gateway in the wall which ran round the castle, suiting itself to the irregularities of the hill. They crossed the courtyard and dismounted. The grooms led their horses away, and, as the big double doors opened, they went in, one of them, standing aside for the younger of his companions but entering before the other. In the great hall whose walls were adorned with horns and heads and tusks, and whose floor was almost completely carpeted with skins, they gave their weapons to a couple of footmen; and as they did so he saw the slim and yet stately figure of a woman coming down the winding stair which led into the hall from a broad gallery running round it. As she reached the bottom of the stairway she threw her head back a little, and held out both her hands towards the man who had come in second. As the light of a great swinging lamp above the stairway fell upon her upturned face, he recognised the Countess Hermia von Zastrow, the reigning European beauty whose portrait in the illustrated papers, and in the great photographer's windows, was almost as familiar as that of Queen Alexandra.

The Count—for the handsome young hunter who now took her hands could now be no other than the Prince of Boravia-Trelitz—raised her right hand in courtly fashion to his lips. The other two bowed low before her, and then she led the way up the stairs.

He saw all this as distinctly as though he had been actually present, and yet none of the party seemed to take the slightest notice of him. But he was getting quite accustomed to miracle-working now, and so he accepted the extraordinary conditions of his visions, or whatever it was, with more interest than astonishment. He followed them up the stairs and along the right hand side of the gallery. The Count opened a door of heavy black oak and stood aside for his Countess to enter. Again the younger of his companions went first, and again he followed; then, as the elder man entered and closed the door, the scene was blotted out as though a sudden darkness had fallen upon his eyes.

"Dear me!" he said, getting up and rubbing his temples with both hands. "If I hadn't had so many extraordinary experiences since my promotion to the plane of N^4, I should probably be a little scared as well. But it is really astonishing how soon the trained intellect gets accustomed to anything—even the eccentricities of the fourth dimensional world. Well, well! I hope that's not the end of the adventure, I was getting quite interested. I suppose this must be in some obscure way the reason why those paragraphs in the *Pall Mall* interested me so strangely."

He walked towards the window, pulled the blind aside and looked out. But instead of his own tree-shaded lawn and the wide expanse of moonlit common beyond which he expected to see, he found himself looking, as it were, through a window from the outside into a great, oak-panelled sleeping chamber, lighted by a huge silver lamp hanging from the middle of the painted and corniced ceiling. Against the middle of the left hand side wall, as he was looking into the room, stood one of the huge, heavily-draped, four-post bedsteads in which the great ones of the earth were wont to take their rest a couple of hundred years ago. The curtains were drawn back on both sides. In the middle of the bed lay Count Zastrow, deathly white, with fast-closed eyes and lips, breathing heavily as the rise and fall of the embroidered sheet and silken coverlet which lay across his chest showed. On the right hand side stood the Countess and the two men whom he had seen before; on the other side stood a tall, strikingly handsome woman, whose dark imperious features seemed strangely at variance with the severely fashioned grey dress and the plainly arranged hair which proclaimed her either a nurse or an upper servant.

He saw the elder of the two men lean over the bed and raise one of the sleeper's eyelids with his thumb. The nurse took up a lighted taper by the table beside her and passed it in front of the opened eye. The man closed the eyelid, and turned and said something to the Countess and the other man. The Countess nodded and smiled, not quite as a man likes to see a woman smile, and, with a swift glance at the motionless figure on the bed, turned away and left the room. The nurse said something to the two men, and as the door closed behind her the scene changed again.

This time he was not looking into a window, but out of one. He was gazing over a vast expanse of forest pierced by a broad, straight road which led for several miles, as it seemed to him, between two dark walls of thickly-growing pines until it ended abruptly with the forest and opened out on a tiny sand-fringed inlet whose narrow mouth was guarded by two little outcrops of rock half a mile to seaward.

A carriage drawn by four black horses rolled rapidly along the road, swung out on to the beach, and stopped. Almost at the same moment a grey-painted, six-oared boat grounded on the sandy beach. A couple of men landed from her, and as the carriage door opened, they saluted. The Count's two guests got out and the others entered the carriage, then one of them got out again followed by the other, and between them they carried a limp, motionless human form completely covered by a great rug of dark fur. It was taken to the boat. All embarked, and the pinnace shot away out through the little headlands. A mile out to seaward lay the long black shape of a torpedo destroyer. The pinnace ran alongside and they all went on board, two of the sailors carrying the body as before.

Professor Marmion found himself accompanying them. The body was taken into a little cabin and laid in a berth. The rug was turned down from the face, and he recognised Prince Zastrow. A few minutes later he found himself in the main cabin of the destroyer. The two men who had come in the carriage were sitting at a little table with a man in mufti. This man raised his head and said something. He did not hear the words—but, to his amazement, he recognised the handsome face as that of Prince Oscarovitch, whom he had never seen before he came as his guest to the garden-party at "The Wilderness."

On the bulkhead of the cabin at the Prince's head there hung a little block-calendar, and the exposed leaf showed the date, Monday, 6th June. As he read it an impulse caused him to look round at the calendar standing upon his own mantel-shelf. It showed the date, Friday, 24th June. He turned back to the window and saw nothing but his own lawn and the moonlit Common beyond.

CHAPTER XVII
M. NICOL HENDRY

Franklin Marmion sat down and began to think the situation over. It was not an easy one, for, as it appeared to him, it would be very difficult, if not impossible, for Nitocris and himself to help in the elucidation of the Zastrow mystery, and the prevention of any European complications that might arise out of it, on both the higher and the lower planes of existence. Of course, it would have been perfectly easy to do so in one sense, for now, practically nothing in human affairs was impossible of achievement to them; but, on the other hand, it would never do to allow people on the lower plane to become aware of their extra-human powers. This was out of the question for many reasons, not the least of which was that they had their lives to live under the ordinary conditions of time and space and among their fellow-mortals, every one of whom would shun them in fear, perhaps even horror, if they knew their secret. What, for instance, would happen to Nitocris in her temporal state if even only Merrill came to know it? No, the idea was certainly beyond the possibility of consideration.

At the same time, it was to some extent necessary that they should work on both planes if they were to reap the full advantage of their recently acquired powers, and out of this dilemma there appeared to be only one way open to the Professor: he must have the assistance of others to do on the lower plane the work that he would, as it were, direct from the higher. The question was, who? Obviously it must be some one upon whose discretion absolute reliance could be placed. He must be highly skilled in police work, and have a reputation to enhance or lose as the result might decide. Suddenly a name occurred to him. A short time ago his friend the President had been telling him the inner story of a very intricate case which had involved a scandal of two Courts. Only the most meagre details had obviously been permitted to appear in the papers, but His Lordship had told him that it had been solved and settled almost entirely by the skill and diplomacy of a M. Nicol Hendry, who held the little advertised but highly responsible position of Head of the English Department of the International Police Bureau.

"That's the very man," he said, "the very man, and I shouldn't wonder if he's engaged on this particular case. It's too late to wire, and, besides, that would look suspicious. I could telephone to Scotland Yard, but I don't want even the police to know I want him until I've seen him. No, I'll write a note: it will go by the early post, and no one will know where it comes from."

Just as lunch was over the next day the front door bell tingled, and presently the parlour-maid knocked, and came in with a card on a silver salver:

"I have shown the gentleman into the drawing-room, sir. He says that he has an appointment with you for half-past two."

"Very well: I will be up in a moment, Annie." Then, as she closed the door, he gave Nitocris the card, and continued: "Our ally on the lower plane that may be. You say you wouldn't care to be present and help me with your opinion?"

"Oh no, Dad. I don't want any one to know that I am taking any part in this little adventure. But if you will introduce him afterwards, I'll tell you what I think. You know, women generally judge other people that way."

"Very well," laughed her father, as he turned to the door, "that will be best. If everything goes right and I think I can work with him, I shall bring him upstairs and you can give him a cup of tea. If I don't, you will know that he won't do."

"Good-bye, then, for the present," she smiled, "and don't frighten the poor man, if you can help it. I dare say he's only an exaggerated policeman, after all."

But it was a very different sort of person whom Franklin Marmion greeted in the drawing-room. M. Nicol Hendry was a slimly but strongly-built man of about forty. His high, somewhat narrow forehead was framed with close-cut, crinkly, reddish-brown hair. Under well-defined brown eyebrows shone a pair of alert steel-grey eyes of almost startling brilliancy. His nose was a trifle long and slightly aquiline. A carefully-trained golden-brown moustache half-concealed firm, thinly-cut lips, and a closely-trimmed, pointed beard just revealed the strength of the chin beneath. He was dressed in a dark grey frock-coat suit, and wore a pinky-red wild rose, which he had plucked on the Common, in his button-hole. As he shook hands with him the Professor made a mental note of him as an embodiment of strength, keenness, and quiet inflexibility: a summing-up which was pretty near the truth.

"Good afternoon, M. Hendry," he said, as the hands and eyes met.

"Good afternoon, Professor," returned the other in a gentle voice, and almost perfect English. "May I ask to what happy circumstance—at least, I hope it is a happy one—I owe the honour of making the acquaintance of the gentleman who has succeeded in mystifying all the mathematicians of Europe?"

"Well," said Franklin Marmion with a smile, "I don't know whether there is so very much honour about that, but I do know that your time is very valuable and that I have already taken up a good deal of it by bringing you all the way out here, so I will come to the point at once. But wait a moment. Come down into my study. We can talk more comfortably there." When the Professor had given his guest a cigar and lit his pipe, he said quite abruptly: "It is about the Zastrow affair."

If he had said it was about the last Grand Ducal plot in the Peterhof, M. Hendry could not have been inwardly more astonished. Outwardly the Professor might have mentioned the last commonplace murder. Only his eyelids lifted a little as he replied:

"Ah, indeed? Well, really, Professor, you must forgive me for saying that that is about the very last matter I should have expected you to have brought up. All the world knows you as one of its most distinguished men of science, now, of course, more distinguished than ever; but I hardly think any one would have expected you to interest yourself in political mysteries. I have a recollection of hearing or reading somewhere that politics were your pet aversion."

"So they are," replied Franklin Marmion, with a short laugh. "I consider ordinary politics—juggling with phrases to delude the ignorance and flatter the prejudices of the mob, and bartering principles for place and power—to be about the most contemptible vocation a man can descend to, but those are low politics in more senses than one. Now high politics, as a psychological study, to an outsider are a very different matter. But I am digressing. I did not invite you here to discuss trivialities like these. I want to ask you—of course, you will not answer me unless you like—whether you are connected, professionally or otherwise, with the Zastrow affair?"

M. Hendry looked down at the toes of his perfectly-shaped boots for a moment or two. Then he raised his head and said good-humouredly:

"Professor, I know that there is no more honourable man in the world than you, but even from you I must ask frankly your reasons for asking that question?"

"You have a perfect right to do that, my dear sir," was the quiet reply. "If you say 'yes,' I am anxious to help you: if you say 'no,' I should like you to help me: if you don't care to answer, there is an end of the matter. Those are my reasons."

It took a good deal to astonish Nicol Hendry, but he was considerably astonished now. Yet it was impossible to have the remotest doubt of Franklin Marmion's absolute earnestness. But why should he of all men on earth want to unravel the Zastrow mystery? What interest save the merest curiosity could he have in the matter? And yet he was by no means the sort of man to be merely curious. The very strangeness of his proposition half-convinced him that there must be some other very strong reason underlying those which he had given. Again, he was to be perfectly trusted, so no harm could be done trying to discover if this was so, since if he could help he would do so loyally. So he told him.

"Yes, Professor," he said, looking keenly into his eyes, "I am interested in the *affaire*, professionally interested, and, I may add, very deeply interested, to boot."

"I am glad to hear that," said Franklin Marmion with unexpected earnestness. "Now, the next question is: Will you accept my assistance, whatever it may be, under my own conditions, which are these: No one but yourself shall know that I am helping you, and you yourself will not ask me how I help you."

Once more a puzzle. Nicol Hendry thought for a few seconds before he replied slowly:

"Yes, Professor. As long as you do help us I don't care either why or how, for, as I may now be quite frank with you, we certainly want help of some sort very badly. The papers are quite right for once. Neither here nor on the Continent have we found a single clue worth picking up. It is humiliating, but it is true."

"Then before you go I hope I shall be able to give you some that will be worth picking up, and keeping too," said the scientist with a faint smile; "at any rate, I think I can put you upon certain lines of enquiry which you will find it profitable to trace out."

Nicol Hendry was an ambitious man, and he would have given a good deal to have known what was passing in the other's mind just then, but his expression betrayed nothing more than interested anticipation.

"We shall be entirely grateful to you if you will, Professor," he murmured.

"I have no doubt of that, my dear sir. Now, to begin with: I presume that there are photographs of the persons mentioned in the newspapers as being in the Castle of Trelitz with the Prince on the last day that he was known to be there?"

"Certainly; we should scarcely leave a simple preliminary like that neglected," smiled Nicol Hendry. "With the exception of the Fraülein Hulda von Tyssen, the Princess' Lady of the Bedchamber, all have been photographed for publication, and hers we have got through a private source. The Chief of each of our Departments has a copy of them, and I happen to have mine in my pocket now, if you would like to see them. The Princess, of course, you must have seen. She is in every photographer's window in the West End."

"Oh yes, I have seen her. Who has not? She is a singularly beautiful woman. But I should very much like to see the others, if I may."

The Chef de Bureau looked at him sharply as he took a small square morocco case out of his inner pocket and opened it. Going to a little table he spread out five small unmounted photographs upon it. He put two of them on one side, saying:

"Those, of course, you know; they are the Prince and Princess. This one is Count Ulik von Kessner, High Chamberlain of Boravia; this, Captain Alexis Vollmar; and this is Fraülein von Tyssen."

Franklin Marmion looked at them with much more than ordinary interest, for he recognised all five as clearly as though he had just left them in his own dining-room.

"There are no suspicions attaching to any of these people, I suppose?" he said carelessly.

"My dear Professor," replied Nicol Hendry a little coldly, "those who write stories about our profession always say that it is our invariable rule to suspect everybody, but we have a little common-sense, and we know the records of these ladies and gentlemen in the minutest detail from the Prince himself to Fraülein Hulda. We have not the slightest reason to suspect any of them."

"Ah, just so," said the other musingly; "no, of course you wouldn't have, and, unfortunately, I cannot tell you why you should. But I'll tell you this: if you ever do find cause to suspect any of these persons, you will find that this group is not complete. It ought to contain the photograph of Prince Oscar Oscarovitch."

"Prince Oscar Oscarovitch!" exclaimed Nicol Hendry, staring at him this time with wide-open eyes. "Why on earth should you——"

"Pardon me, my dear sir," interrupted Franklin Marmion gently, "remember that you are not supposed to care anything about the why or the how. I have already explained that I cannot explain."

"A thousand pardons, Professor. I don't often forget myself, but I did then. You took me so utterly by surprise."

"I fancy that you will be a good deal more surprised before you have come to the end of this affair," was the smiling but almost exasperating reply; "but, as I implied, I can only give you clues. I cannot even tell you how I get them, and it is for you to follow them or not as your judgment dictates. Now, here are one or two to go on with. Try and find out whether or not there was a four-funnelled Russian destroyer anywhere in the neighbourhood of Trelitz on the night of the 6th. Trace as closely as you can the movements of Prince Oscarovitch on that and the two preceding days. Try and find out whether or not a large closed chariot something like a barouche, drawn by four black horses, went from anywhere in the direction of the Castle on that day. And lastly, keep a very close eye upon the Egyptian Adept, as he calls himself—his name is Phadrig Amena—who worked those alleged miracles at my daughter's garden-party the other day. The Prince practically invited himself, and brought this fellow with him. If you can find out the true relationship between them I think you will have found out enough to keep you rather busy for the present. If you do think anything of these little points and examine them, let me know how you get on. We are going abroad for a bit of a holiday, but I will send you my address every now and then. Now, let us go back into the drawing-room, and my daughter will give us some tea."

When Nicol Hendry left "The Wilderness" that afternoon he was about the most mystified man in London. After he had gone, Franklin Marmion said to Nitocris:

"Well, Niti, what do you think of our gimlet-eyed friend? Will he do?"

"Yes, Dad; I like his manner, and he seems very clever in his own way. Quite a gentleman, too," she replied.

"I'm glad you think that," he added; "but what a pity it is that we could not get the world to accept fourth dimensional evidence without turning the said world inside out. We could clear up the whole *affaire* Zastrow in a week then."

"But we shouldn't enjoy our holiday as much, I'm afraid, it would be too exciting," concluded Nitocris.

CHAPTER XVIII
MURDER BY SUGGESTION

Two days later the Marmions left London for Copenhagen, whence they intended to take a trip among the Baltic Islands, now looking their brightest and prettiest, then up along the Norwegian Fiords, just before the tourist rush began, and finally across from Trondjem to Iceland. They were both excellent sailors, and both disliked crowds, especially when the said crowds were pleasure-hunting. Moreover, they had now a particular reason for being alone that they might enjoy together—they, the only two mortals who could do so—the countless marvels of that new existence which had now become possible for them. Where, too, could they do this to more advantage than in the ancient Northland, whose marvellous past would now be to them even as the present of their own temporal lives?

The Van Huysmans, and, of course, Lord Lester Leighton, were to remain in London until the end of the Season. Uncle Ephraim had cabled warm congratulations and large credits, and so Brenda, very naturally as a newly-engaged girl and a prospective Countess, wanted all that London and Ranelagh and Henley, Ascot and Goodwood and Cowes, could give her before her devoted lover's yacht carried them off to the Mediterranean. Later in the autumn they were all to go over to the States to spend the winter in Washington and New York, whence they were to return to London for the wedding in May: surely as pleasant a programme—I fear that Miss Brenda spelt it "program"—as could be desired even by a fair maiden upon whom the kindly Fates had already showered their choicest gifts. The only bitter drop in the family cup of content was the fact that Professor van Huysman was as far away as ever from the exposure of the fallacy which, as he was immovably convinced, those abominable demonstrations *must* contain.

After due consultation between Nicol Hendry and his colleagues of France, Germany, and Russia, it was decided to follow up the clues which he had so mysteriously received. The others would, of course, have been very glad to know where and how he got them, but at the outset he had put them on their honour not to ask, and so professional etiquette made it impossible for them to do anything but accept his assurance that he had received them from a source which was quite beyond reproach. Once they accepted the situation, they got to work with a quiet thoroughness which resulted in the spreading of an invisible but unbreakable net round the footsteps of every one of the suspects from the great Oscarovitch himself to the humble seller of curios in Candler's Court, and his still humbler friends Pent-Ah and Neb-Anat, who were known to the few who knew them as Mr and Mrs Pentana, renovators, and, possibly manufacturers, of ancient gems and relics.

But to one pair of eyes, at least, the police-net was as plainly visible as a spider's web hanging in the sunlight.

Within three days Phadrig received a visit from a shabbily-dressed but well-to-do Jew trader with whom he had done business before, who wanted to know if he could put him in the way of getting some really good old Egyptian gems and jewellery to show on approval to a wealthy patron who wanted to give his daughter a set of rare and uncommon ornaments on her wedding day. It was by this means, by acting as an intermediary between those who had something to sell and those who wished to buy, that Phadrig was supposed to make his modest living. His knowledge of Eastern antiquities was admittedly great, though, of course, no one knew how great, and he had often been asked why, instead of living in such a wretched way, he did not start a little business for himself; to which he always replied that he had no capital, and that he preferred independence, however poor, to the cares and ties of regular trading.

When the Jew had stated his business, Phadrig looked at him with sleepy eyes with a strange expression in them which, for some reason or other, held his visitor's usually shifty gaze fixed, and said in a slow, gentle voice:

"It is very kind of you, Mr Josephus, to bring me all these nice little commissions. They are of much benefit to a poor student of antiquities like myself, although I do not like trading in things that I love. Still, one must live if one would study. Now, I had a gem sent to me the other day which I would dearly love to possess, but, alas! as well might I long for the Koh-i-Noor itself. Moreover, it is already promised—nay, as good as sold. But what have the poor to do with such splendours save to help the rich to buy them!"

The Jew's prominent eyes shone with an inward light at the mention of the gem, and he said in a coaxing voice:

"My dear Phadrig, we have always been friends for ever so long, and you say I've been a good customer to you. Might I have a look at that gem? You know how fond I am of the pretty things. Have you got it here?"

"Yes, and you shall see it with pleasure, my good Josephus," replied Phadrig, well knowing the thought that was in his mind when he asked if he had the gem there in that shabby, unprotected room.

He went to the old oak secretaire, unlocked a cupboard at the side, and then a drawer within it, followed in every motion by the gleaming eyes of the Jew, and took from it a leather parcel. He undid this and produced a box, about four inches long and three wide, of plain black polished wood. It looked solid, but Phadrig made a swift motion with his fingers, and one half of it slid off the other. He held it towards his visitor, and said:

"What do you think of that as a specimen of ancient art, Mr Josephus?"

The Jew looked. The inside of the box seemed filled with green light tinted with yellow. Out of the midst of it began to shine a deeper green light which crystallised into the most glorious emerald that he had ever even dreamt of. It was fully an inch square, flawless, and of perfect colour. The yellow sheen came from a framework of heavy, exquisitely-wrought gold. Phadrig took it out and held it before him, and the green light seemed to radiate through the dull atmosphere of the room. The Jew stared at it with bulging eyes and trembling under-lip, and his hands went out towards it with a gesture which seemed like worship.

"God of Israel," he gasped, "was anything so splendid ever seen before! Mr Phadrig, is it—is it real?"

"Real?" echoed the Egyptian scornfully. "Did you ever see light like that come out of a sham stone? You should know more about gems than that, Mr Josephus."

"Ah yes, yes, of course. It is glorious; it is worthy to shine on the breastplate of the High Priest—and what a price it must be! Is it allowed to ask the name of the great millionaire for whom it is destined?"

"Yes. It will in a few hours be the property of Prince Oscar Oscarovitch."

As Phadrig spoke he hid the gem in his hand. His voice was so changed that the Jew looked up at him. His eyes were wide open now, and glowing with a fire that made them look almost dull red. They seemed to see right through his eyeballs and look into his brain. Josephus started as though he had been struck. He tried to turn his head away, but the terrible eyes held him. His fat, greasy, olive face grew grey and dry, and his head shook from side to side.

"What is the matter, my dear Mr Josephus?" asked Phadrig, in slow, stern tones. "The mention of the Prince seems to have affected your nerves. Are you acquainted with His Highness?"

"Me? I? Why, how should I know a great man like the noble Prince? No, no; of course I know him as a very grand and great gentleman, but that is all, really all, my dear Phadrig."

"Yes, yes, of course," said the Egyptian, once more in his gentle voice; "would not be likely, would it? Now, if you would like to look at the gem more closely, go and sit down there by the light and take it in your hand. You will see that it is engraved with hieroglyphics. They say that this jewel was once the property of Rameses the Great of Egypt, and was given by him to his daughter Nitocris."

This information did not interest the Jew in the slightest, since he had never heard the names in his life; but the delight and honour of holding such a glorious gem in his hand even for a few minutes was ecstasy to him. He sat down, and held out his fat, trembling hand greedily. With a smile of contempt Phadrig placed the jewel in it, and said:

"Examine it closely, my friend. It is well worth it, and it may be long before you see another like it."

"Like—like *it*, like *this*! By the beard of Father Moses, I should think not—I should think—I should—oh, beautiful—glor—glorious—splendid—did—splen—oh, what a light—li—light—li—oh——!"

As each of the disjointed syllables came from his shaking lips he mumbled more and more, and his head sank lower towards the priceless thing in his palm. As he gazed, the stone grew round and bigger and brighter, till it seemed like a great green-blazing eye glaring into the utmost depths of his being. Then the light suddenly went out, his head fell on his breast, and as his hand sank, Phadrig caught it and took away the jewel. Then he put the Jew back in the chair, and standing in front of him began in a slow, penetrating voice:

"Isaac Josephus, thou hast gazed upon the Horus Stone, and he who doeth that may not answer the questions of an Adept with lies save at the price of his life. Now answer me truly, or to-morrow morning those of thine household shall find thee dead in thy bed."

Wide open the eyes of the hypnotised man stared at him, and the loose lips quivered, but these were the only signs of life.

"Thou art not only a dealer in gems and curious things: thou art also a spy of the police; is not that so?"

"Yes."

"Believing that I am a very poor man, yet knowing that I dealt with objects of value, they thought me to be one who receives such things from thieves to sell them again, since they could not. Is that so?"

"Yes."

"And, believing this, and knowing thee to have dealings with me, they bribed thee to come here as my friend and fellow-dealer and spy upon my actions, so that they might have evidence against me and cast me into prison. Is that so?"

"Yes."

"Late on the last night but one thou didst go to the house of Nicol Hendry, who is no common catcher of thieves, but a spy of nations whose business is with the great ones of the earth. Tell me: whom did thy business with him concern?"

"Prince Oscarovitch and yourself."

"What were his orders?"

"To watch you both, especially you, and find out when you went to him, and why you were sometimes a poor devil in a miserable hole like this, and sometimes a swell going to swagger places with him."

"How were you going to do this?"

"I know your servant or chum, Mr Pentana. I've lent him money: and Peter Petroff, the Prince's particular servant, gambles like a lord, and he owes me and a friend of mine a lot of money. We were going to work through them."

"It is enough; and well for you that you have answered truthfully. Now tell me: do you know how to use a revolver?"

"Never fired a shot in my life."

Phadrig went to the secretaire and took a common, cheap revolver, identical with thousands of others which our criminally careless Government allows to be bought every day without the production of a licence—just a hooligan's weapon, in fact—went back and put it into the Jew's hand. He raised the hand several times, and pointed the muzzle to the temple, keeping the forefinger on the trigger. At length he let go the wrist, and said in a gentle, persuading tone:

"That is the way to handle a revolver when you are going to shoot, my dear Josephus. Now, let me see if you can do it by yourself."

With mechanical precision the Jew's arm went up until the muzzle touched his temple. Again and again he did the same thing at Phadrig's bidding, till at length he said rather more peremptorily:

"Now pull the trigger!"

The finger tightened and the hammer clicked. Five times more was the operation repeated, and then Phadrig gently took the revolver and laid the hand down. He went to the secretaire and loaded the six chambers, cocked the weapon and put it into the right hand side-pocket of the lounge jacket which Josephus was wearing, and said deliberately:

"Now remember, my dear Josephus: you will go straight back to your office in Waterloo Road and let yourself in with your key. In your private room you will see a man who wants to rob you of some valuable papers. You will be ruined if he gets them, so you must take your pistol out of your pocket and shoot him. Do you quite understand me?"

"Yes; I am to shoot him."

"That is right. Now, if you do not go he will have them before you get there. Get up and we will say good-night. You must not put your hand in your pocket until you see the man who wants to rob you. Good-night. There is your hat."

"Good-night!"

Mr Isaac Josephus put on his hat and walked away to his death with the motions of a mechanical doll.

CHAPTER XIX
THE HORUS STONE

An hour later Phadrig, the poor curio dealer, had disappeared, and Mr Phadrig Amena, the wonder-working Adept, clad in evening clothes and a light overcoat, alighted from a hansom at the great entrance to the Royal Court Mansions. The huge, gorgeously uniformed guardian of the Gilded Gates was saluting at his elbow in an instant, for a friend of Princes is a very great man in the eyes of even such dignitaries as he.

"The Prince expects you, sir," he said, loud enough to make the title heard by those who were standing by. "Will you be good enough to walk in? I will discharge the cab."

He stood aside with a bow and another salute, and Phadrig walked lightly up the broad steps. Peter Petroff opened the door of the flat, bowing low, and conducted him to his master's sanctum. Evidently he was expected, for the coffee apparatus stood ready on the Moorish table beside the cosy chair which he was wont to occupy. The Prince, who was standing on a white bear's skin by the mantel, motioned him to it, saying:

"Ah, Phadrig, my friend, punctual, of course; and equally, of course, you have something important to impart. Your wire just caught me in time to put off an engagement which, happily, is of no great consequence. There's the coffee, and you'll find the cigars you like in the second drawer. Now, what is the news?"

His guest filled a cup of coffee and took a cigar and lit it before he replied. Then, turning to the Prince, he said in his usual slow, even tone:

"Highness, I regret to say that my news is both urgent and bad."

"It would naturally be urgent," said the Prince, turning quickly towards him, "but bad I hardly expected. Well, all news cannot be good. What is it?"

"I fear that my warning was even more urgent than I thought it myself—I mean, in point of time. Your Highness is already being watched."

"What! A Prince of the Empire, the man whom they call the Modern Skobeleff, an intimate of Nicholas! What should I be watched for?" exclaimed the Prince, half angry and half astonished. "The thing is ridiculous; another of your dreams!"

"Ridiculous it may be, Highness," replied Phadrig, quite unruffled, "but it is no dream; and, moreover, the eyes which are watching you are keen ones—and they are everywhere. You are under the surveillance of the International Police."

These were not words which even a Prince of the Holy Russian Empire cared to hear. Oscarovitch was silent for a few moments, for the earnestness, and yet the calmness, with which they were spoken made it impossible for him to doubt them. As he had asked, what could such a man as he be watched for by this thousand-eyed organisation of which he himself was one of the supreme Directors? It was impossible that these people could suspect his great scheme of treachery and self-aggrandisement. That was known to only three persons in the world—himself, Phadrig, and the Princess Hermia; and the Princess, the woman who had willingly sacrificed her brilliant young husband to her guilty love and her boundless ambition—no, she could be no traitress. It must be something else: and yet what?

He took two or three rapid turns up and down the room, chewing and puffing at his cigar, until he stopped before Phadrig, and said quietly, but with angry eyes:

"Very well, we will grant that I am watched by the International. Tell me how you came to know it."

The Egyptian took a few sips of his coffee, and then related almost word for word his interview with Josephus. He ended by saying:

"Your Highness may believe or not now as you please, but I presume you will when you read in your paper to-morrow morning of the suicide of a respectable Hebrew merchant named Isaac Josephus at the address which I have mentioned."

Oscarovitch had pretty strong nerves, and he was well accustomed to regard any kind of crime as a quite proper means of furthering political ends: but there was something in this man's utter soullessness and the weird horror of the crime which he had just accomplished—for by this time his victim would be already lying self-slain on the floor of his own spider's lair—that chilled him, cold-blooded as he was. He looked at him lounging in his chair and calmly puffing the smoke from his half-smiling lips as though he hadn't a thought beyond the little blue rings that he was making.

"That was a devilish thing to do, Phadrig!" he said, a little above a whisper.

"Devilish, possibly, Highness, but necessary, of a certainty," was the quiet reply. "You will agree with me that Nicol Hendry is a dangerous antagonist even for you, and as for me—no doubt he thinks that he can crush me under his foot whenever he chooses to put it down. I should like to know his feelings as he reads of his spy's suicide when he had only just got to work."

"It will certainly be somewhat of a shock to him and his colleagues, and for that reason I am inclined, on second thoughts, to agree that it was necessary, and ghastly, as I confess; it seems to me, I think, that you took the best means to give them a salutary warning. After all, the life of an individual, and that individual a Jew, does not count for much when the fate of empires is at stake. What puzzles me is how these fellows came to suspect me, and what do they suspect me of. I suppose you have no idea on the subject, have you?"

He looked at him keenly as he spoke, but he might as well have looked at the face of a graven image. Then, like a flash of inspiration, the Zastrow affair leapt into his mind. Had his connection with that, by any extraordinary chance, come to the knowledge of the International? The thought was distinctly disquieting. Phadrig had helped in this with his strange arts. He would discuss this phase of the matter with him afterwards.

Phadrig replied, returning his glance:

"Highness, I have only one explanation to offer, and that you have already refused. Were I to speak of any other it would only be vain invention."

"You mean about Professor Marmion and his mathematical miracles?" said the Prince somewhat uneasily.

"I do," replied the Egyptian firmly. "I say now what I thought when I saw him work them. I did not believe that any man could have done what he did unless he had attained to what we styled in the ancient days the Perfect Knowledge, or, as they term it to-day, passed the border between the states of three and four dimensions. If Professor Marmion has achieved that triumph of virtue and intelligence—and in the days that I can remember there were more than one of the adepts who had done so—then Your Highness's Imperial designs must be as well known to him as to yourself: nay, better, for, while you can see only a part, the beginning and a little way beyond, he can see the whole, even to the end; for in that state, as we were taught, past, present, and future are one. Now, only three persons know of the project, and treason among them is not within the limits of reason, wherefore I would again ask Your Highness to believe that such information as the International may have has been given them directly or indirectly by Professor Marmion."

"But," said the Prince, who was now evidently wavering in his scepticism, since Phadrig's explanation of the mystery really seemed to be the only feasible one, impossible as it looked to him, "granted all you say, what possible interest could Professor Marmion, whether he's living in this world or the one of four dimensions, have in interfering in such a project, even if he did know all about it, especially as every educated Englishman admits that the state of affairs in Russia could hardly be worse than it is? I cannot see what conceivable interest he can have in the matter."

"But, Highness, his interest may be a private and not a public one."

"What do you mean by that, Phadrig?" asked the Prince sharply.

"As I have said," replied the Egyptian slowly, "it may be that his daughter, who was once the Queen, has also attained to the Knowledge. In that case the love which Your Highness so suddenly conceived for her would instantly bring you within the sphere of his and her influence and power. Now, she, as Nitocris Marmion, the mortal, is betrothed to the English officer, Merrill. She loves him, and therefore, since you are great and powerful in the earth-life, your ruin, or even your death, might seem necessary to remove you from her path."

Oscarovitch shivered in spite of all his courage and self-control. The idea of fearing anything human had never occurred to him after his first battle; but this, if true, was a very different matter. To be threatened with ruin or death by a power which he could not even see, to contend against enemies who could read his very thoughts, and even be present in a room with him without his knowing it—as Phadrig had assured him more than once that they could be—was totally beyond the power of the bravest or strongest of men. No, it was impossible: he could not, would not, believe that, such a thing could be. His invincible materialism came suddenly to his aid, and saved him from the reproach of fear in his own eyes.

"No, Phadrig," he said, with a gesture of impatience, "that is not to be credited. To you it may seem a reality: to me it can never be anything more than a phantasy of intellect run mad on a single point—which, I need hardly remind you, is a by no means uncommon failing of the greatest of minds. Another reason has just occurred to me which would need no such fantastic explanation."

"And that, Highness?" queried Phadrig, looking up with an almost imperceptible shrug of his shoulders.

"The Zastrow affair. Unlikely as it seems, it is not impossible that there has been treason there. I have many enemies in both Russia and Germany, and it is well known that Zastrow and I were rivals once. Yes, that is it: it must be so, and therefore we must prepare to fight the International; and with such weapons as you are able to use there is not much reason why we should fear them."

He dismissed the subject with an imperious wave of his hand, and continued in an altered tone:

"And now, *àpropos* of your weapons. Tell me something about this wonderful gem with which you hypnotised the Jew."

"I will not only tell you about it, Highness, I will show it to you, if you desire to see it," replied Phadrig, who now fully recognised the hopelessness of overcoming the blind materialism which was, of course, inevitable to the life-condition in which the Prince had his present being.

"What! you have brought it with you! Excellent! Now I think we shall be able to talk on pleasanter subjects than conspiracies and such phantasms as the Fourth Dimension!" exclaimed Oscarovitch, who, like all Russians, was almost passionately fond of gems. "Fancy asking a Russian if he desires to see such a thing as that!"

"Your Excellency must be careful not to look at it too long or closely," said Phadrig, putting his hand down inside his waistcoat and drawing out a wash-leather bag. "As I have told you, it possesses certain qualities which are not to be trifled with. You are, of course, aware that many Eastern gems are credited with hypnotic powers. This one undoubtedly has them."

As he spoke he drew out the emerald, and held it by the clasp under a cluster of electric lights.

"What a glorious gem!" exclaimed the Prince, starting forward to look at it more closely. "There is nothing to compare with it even among the Imperial jewels of Russia."

"Have a care, Highness," said the Egyptian, raising his left hand, "unless you wish to fall under its influence. Once it seized your gaze you could not withdraw it without the permission of its possessor, and meanwhile he would have complete mastery of you. I am your faithful servant, and therefore I warn you."

Was there just the faintest suspicion of a sneer in his voice as he said this? If there was, Oscarovitch did not notice it. He was already too much under the charm of the Horus Stone. Phadrig suddenly put his hand over the gem and went on. "The story of this jewel, Highness, is that many ages ago, before the beginning of the First Dynasty, a little raft of a strange wood, as white as ivory and shaped like a river-lily, came floating down the Nile at full flood-time and drifted to the shore in front of the house of a wise and holy man who was reputed to hold perpetual communion with the gods. On the raft was a cradle of white wicker-work lined with down, upon which lay a man-child of such exquisite beauty that he could scarce have been born of mortal parents. His body was bare, but round his neck was a glistening chain of marvellously wrought gold, fastened to which was this gem lying on his breast. This was doubtless the origin of the Hebrew fable of the finding of Moses, who, as all scholars know, was not a Hebrew, but an Egyptian priest in the House of Ra.

"The holy man took him into his home, burying the chain and gem, lest it might bring temptation to those who saw them; and as the boy grew to manhood he taught him all his lore, until he, too, was wise enough to be admitted into the communion of the gods, which afterwards was called by the adepts the Perfect Knowledge. On the gem are engraved the three symbols by which the Trinity—Osiris, Isis, and Horus; Father: Mother, and Child, the antetype of Humanity—became known and worshipped. The holy man divined that the boy was the incarnation of Horus sent thus to earth to teach men the way of knowledge, which is the only righteousness, since those who know all cannot sin. Where his house stood was built the first Temple of the Divine Trinity, and of this Horus became High Priest. He crowned the King in the land, and hung this gem round his neck as the symbol of his kingship and the approval of the gods.

"From the first king it was handed down from monarch to monarch through all the changes of dynasties, until it hung from the royal chain of the great Rameses; and by him it was given to his daughter Nitocris, thereby making her Queen of Egypt after him; and she wore it on that fatal night of the death-bridal when, rather than wed with you, who were then Menkau-Ra, Lord of War, she flooded the banqueting hall of Pepi and drowned herself and all her guests—which, Highness, is an omen that it were well for you not to forget should you persist in your pursuit of the daughter of Professor Marmion."

Oscarovitch was a man of vivid imagination, as all great soldiers and statesmen must be, and so the story of the Horus Stone appealed strongly to him; but what interested him perhaps even more was the spectacle of this man, who had just been guilty of a peculiarly ghastly form of murder, sitting there and telling with simple eloquence and evident reverence the sacred Myth out of which what was perhaps the most ancient religion in the world had evolved. He heard him with a silence of both interest and respect until his last sentence. Then he got up and stretched his arms out and said with a laugh:

"Omen, Phadrig! Your tale of the stone has interested me deeply, but I believe no more in the omen than I do in the story. Ay, and even if I did, I would dare all the omens that wizards ever invented for their own profit in trying to make Nitocris Marmion what I want her to be, and what she shall be unless she is the cause of my first failure to achieve what I had set my heart upon. But you have not finished your story. Tell me now how the stone came into your possession, seeing that it was swept out into the Nile hanging on the breast of the Royal Nitocris."

"The next season of Flood, so the records ran, Highness, the skeleton of a woman was washed up to the foot of the river stairs of the House of Ptah, and the stone and chain were found among the weeds which filled the cavity of the chest. They were taken with all reverence to the High Priest, who bore them to the Pharaoh, and, amidst great rejoicing, hung them round his neck. Then from Pharaoh to Pharaoh it came down through the centuries until it fell into the possession of her who wrought the ruin of the Ancient Land. She gave the stone to her lover, and from his body it was taken by a priest of the Ancient Faith who once was Anemen-Ha, and is now Phadrig Amena, the degenerate worker of mean marvels which the ignorant of these days would call miracles did they not take them for conjuring tricks.

"Since then it remained hidden, seen only by the successors of him who rescued it from the plunderers of the body of Antony, until, seemingly in the way of trade, yet doubtless for some deep reason which is not revealed to me, it came back into my hands again. Such so far, Highness, is the end of the story of the Stone of Horus."

"And doubtless more yet remains to be written or told," said the Prince seriously, for he was really impressed in spite of his scepticism. Then, after a little pause, he continued: "Phadrig, you have said that the stone is dangerous to any but its possessor. I wish to possess it. Name your price, and, to half my fortune, you shall have it."

"The stone, Highness," replied the Egyptian, with the shadow of a smile flickering across his lips, "never has been, and never can be, sold for money, so I could not sell it, even if money had value for me, which it has not. There is only one price for it."

"And what is that?"

"A human life—perchance many lives—but all to be paid in succession by him or her who buys it, unless he or she shall attain to the Perfect Knowledge."

"Give it to me, then!" exclaimed Oscarovitch, holding out his hand. "The life I have I will gladly pay for it in the hope of laying it on the breast of the living Nitocris. As I do not believe in any others, I will throw them in. Give it to me!"

"It is a perilous possession, Highness, for one who has not even attained to the Greater Knowledge, as I have. Let me warn you to think again, for once you take it from me the price must be paid to the uttermost pang of the doom that it may bring with it."

"I care nothing about your knowledges, Phadrig," laughed the Prince, still holding out his hand. "It is enough for me to know that it is the most glorious gem on earth, and that it shall help me to win the divinest woman on earth. So, once more, give it to me!"

"Take it, then, Highness," said the Egyptian, with a ring of solemnity in his voice. "Take, and with it all that the High Gods may have in store for you!"

He dropped the more than priceless gem into his hand with as little reluctance as he would have given him a brass trinket. Then he turned away to take another cigar, leaving Oscarovitch gazing in silent ecstasy at, as he thought, his easily-come-by treasure. Then the Prince went to a large panel picture fixed to the wall on the left-hand side of the fireplace, touched it with his finger, and it swung aside, disclosing the door of a small safe built into the wall. He unlocked this, placed the stone in an inner drawer, closed the safe, and put the picture back in its place.

When he sat down again, he said:

"My good friend, I know that it is useless for me to thank you, for even if you wanted thanks I could not do justice to the occasion, as they say in speeches: but I want to ask you just one more question, and then I won't keep you any longer from that delightful Oriental Club of yours which I suppose you are bound to. Now that I have got the stone I am, as you may well believe, more than anxious to find the lady to whom it shall belong—again, as I suppose you would say. To my great disgust, the Professor and his daughter have disappeared from the sphere of London society for a holiday *à deux*, and have, apparently with intent, left all their friends in ignorance of their destination. Have you any idea of it? I know that that Coptic woman whom you employ has been ordered to keep a sharp watch on the movements of Miss Nitocris."

"Yes, Highness," replied Phadrig, "and she has obeyed her orders. The day before they left she waylaid that pretty maid of Miss Marmion's on the Common, and told her fortune. Of course, she talked the usual jargon about lovers and letters and going on a journey, and the maid quite innocently let out that she was going with her master and mistress by steamer to Denmark and up the coast of Norway, and then over to Iceland by the passenger steamers, and that she did not like the idea at all, because she knew that she would be very seasick."

"Excellent! the very thing!" exclaimed the Prince. "It couldn't be better if I had arranged it myself. My yacht is down in the Solent waiting for Cowes Week. I'll be afloat to-morrow. Give that woman a ten-pound note from me with my blessing. Now, I shall leave everything else to you. Do what you think fit with regard to our friends of the International. Kill as many of their spies as you can with safety, and make the chiefs believe that they are fighting the Devil himself. And now, good-night."

When Peter Petroff brought him the papers the next morning, the Prince took up the *Telegraph*, and turned to the page devoted to the minor events of the previous day. His eye was almost immediately caught by a paragraph headed:

"SUICIDE IN THE WATERLOO ROAD

"Shortly after seven last evening the passers-by on the eastern side of this thoroughfare were startled by hearing the report of a firearm, apparently coming from the office of Mr Isaac Josephus at 138a. Constable 206 Q., who was on point-duty near the spot, had seen Mr Josephus enter the office with his key only a few minutes before, walking in a rather curious way, and staring straight before him. As the door was locked, the officer thought it his duty to force it. The door of the inner office was also locked, and when this was opened, the unfortunate man was found lying across the desk with a bullet wound in his temple. His right hand still clutched a cheap revolver which was loaded in five chambers. There appears at present to have been no reason for the rash act. Mr Josephus was a broker dealing chiefly in curios and antique jewellery. Although not in a large way of business, his affairs are understood to have been in a prosperous condition. What makes the tragedy all the more strange is the fact that suicide is almost unknown among persons of the Jewish faith."

Oscarovitch felt a little shiver run down his back as he read the commonplace lines. The man who had done this had been in this room with him a few hours before, and one of the means of murder was now in his safe. It would have been just as easy for Phadrig to have caused him to look upon the fatal gem, left a bottle of poison with him, and told him to take it as medicine on going to bed. The only difference would have been that there would have been a very much greater sensation in the papers.

Nicol Hendry was reading the paragraph about the same time. His eyes contracted, and he stroked his beard with slow motions of his hand. The hand was steady, but even his nerves quivered a little. He divined instantly how the suicide-murder had been brought about, and this very fact, coupled with the absolute impossibility of proving anything, made the affair all the more disquieting.

"So that is the sort of thing we've got to fight, is it? I don't like it. Still, it goes far to prove that the Professor was perfectly right when he told me to keep a sharp eye on Mr Phadrig Amena."

CHAPTER XX
THROUGH THE CENTURIES

As they discovered that the sea journey to Copenhagen would be somewhat tedious and uninteresting, and that the steamers were not exactly palatial, Nitocris and her father decided at the last minute to cross to Ostend, spend a day there and go on to Cologne, put in a couple of days more among its venerable and odorous purlieus, and two more at Hamburg, so that, while the present-day inhabitants were asleep, they might, as Nitocris somewhat flippantly put it, take a trip back through the centuries, and watch the great city grow from the little wooden village of the Ubii and the Roman colony of Agrippina into the Hanse Town of the thirteenth century: watch the laying of the first stone of the mighty Dom, the up-rising of the glorious fabric, and the crowning of the last tower in 1880.

During the journey from Hamburg to Copenhagen, Nitocris, reclining comfortably in a corner of their compartment in the long, easily-moving car, entertained herself with a review of these extraordinary experiences from the point of view of her temporal life, and found them not only extraordinary, but also very curious. She had already learnt that the connecting link between the two existences, when once the border had been passed, was Will: but Will of a far more intense and exalted character than that which was necessary as an incentive to action on the lower plane. There was naturally something that seemed extra-human in the mysterious force which was capable of bidding the present-day world vanish like a shadow into either the future or the past, its solid-seeming substance melt away like "the airy fabric of a vision," and summon in an instant, too brief to be measured, the past from the grave where it lay buried beneath the dust of uncounted ages, or the future from the womb of unborn things.

But to her, at least at first, the strangest part of the new revelation was this: When her will had carried her across the confines of the tri-dimensional world, and she saw the centuries marshalled and motionless before her, she felt not the slightest sense of wonder or awe. She was simply a being apart, moving along their ranks and passing them in review, herself unseen and unknown save by that other being who, in this state, was no longer her father or even her friend, but merely a companion endowed with power and intelligence equal to her own. Her human hopes and fears and loves and passions had, as it were, been left behind. The men and things she saw were absolutely real to her, as they had been to the men of other days, or would be in days to come; but she herself was a pure Intelligence which saw and acted and thought with perfect clearness, but with absolutely no feeling save that of intellectual interest.

She saw armies meet in the shock of battle without a thrill of fear or horror; towns and cities roared up to the unheeding heavens in flame and smoke, and left her standing unmoved amidst their ruins; she heard the screams of agony that rang through the torture chambers without a quiver, and watched the long, pale lines of the martyrs to what in the earth-life was called Religion pass to the stake without a quiver of pity or a thrill of disgust. She stood face to face with the great ones of the earth who have graven their names deep upon the tablets of Time without reverence or admiration; and she witnessed the most heroic deeds and the most atrocious crimes with neither respect for the one nor hatred for the other.

Human history was in her eyes merely a logical sequence of necessary events, neither good nor bad in themselves, but only as they were viewed from this standpoint or that, by the oppressor or the oppressed, the slayer or the slain, the robber or the robbed, the governor or the governed. She learned that human emotion is merely a matter of time and space. One century does not feel the loves and hates of another, and the sorrows of Here have no real sympathy with the sufferings of There. Beyond the Border all these were merely matters of intense intellectual interest.

But when she returned to the temporal life the memory of them was marvellous and terrible. Her heart throbbed with pity and burned with righteous anger. Horror seemed to take hold of her soul and shake it with earthquake shudders when she thought that what she had seen but a few time-moments ago had really come to pass; and she longed for the power to show all this to the men and women of her own passing day, and bid them have done with the poor, shadowy images of themselves, which, had they really been gods, would have made of human life something better and happier and nobler than the ghastly tragedy which, as she had seen with her own eyes, it had been. But she knew that such a power was not hers. She, like her father, had, through the toil and strife and stress of many lives of mingled good and evil, knowledge and ignorance, won her way to the Perfect Knowledge; and so she knew that all these poor kings and slaves, conquerors and conquered, torturers and tortured, were all doing the same thing, were all groping their way through the shadows and the night towards the dawn and the light, through the hell of ignorance to the heaven of knowledge.

And now, too, since the Wisdom of the Ages was hers, she saw that over all the vast, weltering swarm of struggling immortals, hung the inevitable decree of silent, impersonal destiny. "As ye live, so shall ye die; as ye end, so shall ye begin again—in knowledge or ignorance, in good or evil, life after life, death after death, world without end."

It was clear to her now why "some are born to honour and some to dishonour": some to happiness and some to misery, each in his or her degree; why the liver of a good life was happy, no matter what his place in the earth-life might be: and why the evil liver, no matter how high he might stand in his own or others' sight, carried the canker of past misdeeds in his heart. Standing, as she now did, in the midway of the present, looking with single gaze on past and future, she saw at once the honest striver after good in his yesterday-life rise to his reward in the life of to-day, and the dishonest rich and powerful sitting in the high places of to-day cast down into the gutterways of to-morrow. Life had ceased to be a riddle to her now.

What with their halts at Ostend, Cologne, and Hamburg, the thirty-three-hour journey lengthened itself out very pleasantly into a week; and so, when the famous city on the Sound was reached, they were as fresh and unfatigued as they were on the morning that they left "The Wilderness." Of course, they put up at the Hôtel d'Angleterre, and here they enjoyed themselves quietly for four days, for of all European capitals, Copenhagen is one of the pleasantest in which to idle a few fine summer days away.

On the evening of the fourth day they were just sitting down to their table by one of the windows overlooking the Oestergade when Nitocris happened to look up towards the door through which the diners were trickling in an irregular stream of well-dressed men and women. For a moment her eyes became fixed. Then she bent her head over the table, and said:

"Dad, there is Prince Oscarovitch. I wonder what he is doing here? He is alone: please go and ask him to join us. I will tell you why afterwards."

They exchanged glances, and the Professor got up and went towards the door, while his daughter got through a considerable amount of hard thinking in a very short time. She was, of course, perfectly conversant with his share in the Zastrow affair, so far as her father had yet gone with it; but she determined that when Copenhagen had gone to sleep that night they would cross the Border and pay a visit to the Castle of Trelitz at the time of the tragedy, and follow it out as far as it had gone.

It has already been shown that on her first meeting with the Prince she conceived an aversion from him which was then inexplicable save by the ordinary theory of natural antipathy: but now she knew that she had been Nitocris, Queen of Egypt, when he was Menkau-Ra, the Lord of War, who would have forced her to wed him by the might and terror of the sword, and the will of a blind and blood-intoxicated populace. She had hated him then even to death, and now she hated him still in life; wherefore she desired to make his closer acquaintance on the earth-plane on which they had met once more after many lives.

As he had been in those far-off days, so he was now, a splendid specimen of aristocratic humanity. Many eyes had followed her as she had walked to her table, but there were more people in the room now, and as the Prince walked towards her beside the famous Professor who had puzzled all the mathematicians of Europe, the whole crowd of guests was looking at nothing but these three.

"This is indeed good fortune, Miss Marmion, and as good as it is unexpected—which, perhaps makes it all the better! Who would have thought of finding you in Copenhagen?" he said, as he bowed low over her hand.

"If there is any reason at all for it, Prince, it is that my father and I always like to take our holidays at irregular times and in unexpected places: by which, I mean places where we do not expect to meet all our acquaintances," she replied, as she sat down. "I think we manage to bore each other quite enough in London, and we like each other all the better when we meet again."

"Is not that rather an ungracious speech, Niti, seeing that one of the said acquaintances has only just chanced to join us?" said the Professor mildly.

"You mean as regards the Prince?" she laughed. "Certainly not. His Highness is hardly an acquaintance—yet. You know we have only had the pleasure of meeting him once: and then, of course, I said *all* our acquaintances. There might be exceptions."

These words, spoken with a quite indescribable charm, were, as he thought, quite the sweetest that Oscarovitch had heard for many a day. It had been perfectly easy for a man with his official influence to trace by telegraph every movement that the Marmions had made after he had guessed that they would travel by either Calais or Ostend. He had wired for his yacht, the *Grashna,* to meet him at Dover, run across to Ostend, found that they had left there for Cologne with through tickets for Copenhagen, again guessed rightly that they would spend a few days there and in Hamburg, and then steam away for the Sound.

The farther north he travelled, the farther he left Phadrig and his phantasies behind, and the nearer he came to the belief that, if he had only a fair chance and the field to himself, as he intended to have, he would not find very much difficulty in convincing Nitocris that there was no comparison at all between the humble naval officer she had left behind to do his work on his dirty little destroyer, and the millionaire Prince who could give her one of the noblest names in Europe and everything that the heart of woman could desire. And now these sweetly-spoken words and the glance which accompanied them, her undisguised pleasure at the chance meeting, and her father's very evident approval of his presence, quickly but finally convinced him that he had come to a perfectly just conclusion.

Of course, there was the memory of another woman, only a little less fair than Nitocris, who had shut herself up yonder in the gloomy Castle of Trelitz, acting the farce of her official sorrow for love of him, and pining for the time when the finding of her betrayed husband's corpse should leave her free, after a decent interval of mock-mourning, to join her lot with his: but what did that matter? Was it not as easy to get rid of a woman as a man? Was not the fatal beauty of the Horus Stone at his command now that he was its possessor for good or evil? A well-arranged suicide might easily be taken by the world as the excusable, if deplorable, result of her mysterious bereavement.

The conversation during dinner naturally turned on ways and means of travelling, and, when the Professor had sketched out their plans, Oscarovitch said with an admirably simulated deference:

"My dear sir, I most sincerely hope that you and Miss Marmion will not think that I am presuming on an acquaintance which, if only a new one now, may perhaps one day be older, if I venture to suggest another way of making your tour. I am an old voyager in these waters, and I can assure you that the steamers, though vastly improved, have not quite reached the standard of the Atlantic liner."

"Oh, but you know, Prince, we didn't expect it," interrupted Nitocris. "Neither my father nor I have the slightest objection to roughing it a little. In fact, that iş half the fun of wandering."

"And slow travelling between stated points, not always of the greatest or any interest, together with the enforced company of a promiscuous crowd of tourists and commercial travellers, who, by the way, are mostly German, and therefore of nature and necessity disagreeable, would about make up the other half," said Oscarovitch, leaning back in his chair with a low laugh. "No, no, my dear Miss Marmion, I am afraid you would not find that the reality quite squared with the anticipation. Now, may I risk the suspicion of presumption and offer an alternative proposition?"

"Why not?" said Nitocris with a smile, and a glance which dazzled him. "I'm sure it is very kind of you to take so much interest in our poor little attempt to get away for a while from the madding crowd who are doing the round of the same stale, weary pleasures that they try so hard to enjoy year after year, and then come back so tired, after all."

"Then," he replied, looking at them alternately, "as I have your permission, I would suggest that, instead of rushing from fixed point to fixed point in crowded steamers and the shackles of Company or Government regulations, you should take possession of a fairly comfortable steam yacht of a little over a thousand tons which will be entirely at your disposal, and will run you from anywhere to anywhere you choose at any speed you like, from five to thirty-five knots an hour, with properly trained servants to attend to you, and, as the advertisements say, 'every possible comfort and convenience.'"

"Which, of course, means that you have got your yacht here, and are so very kind as to ask us to become your guests for a time," said the Professor, with a suspicion of stiffness. "It is more than generous of you, Prince, but really——"

"But really, my dear sir," Oscarovitch interrupted, with a gesture of deprecation, "I can assure you that, so far as I am concerned, there is no kindness, to say nothing of generosity. It is pure selfishness. This is my position. I have managed to escape for a time from the toils of official work and worry, and the almost equally irksome bonds of that form of penal servitude which is called Society. Like you, I have fled overseas, but, unlike you, I have no company but my own, and I have had a great deal too much of that already, though I have only been three days and nights at sea. I have no plans, I have got nothing to do and nowhere to go; and so, if you and Miss Marmion would take pity on my loneliness all the generosity would be on your side. Of course, I cannot presume to ask you to change your plans all at once, but if you will sleep on my proposition and come and lunch with me to-morrow on board the *Grashna* and take a run up the Sound, say, to Elsinore, you may be able to come to a decision."

It was a lovely night, and so they took their coffee and liqueurs, and the two men their smokes on the balcony overlooking the Oestergade, which might be called the Rue de la Paix of Copenhagen, and watched the well-dressed crowds sauntering to and fro past the brilliantly lighted shops; and Nitocris, who seemed to her father to be in singularly high spirits, sent the conversation rippling over all manner of subjects with the exception of politics and the Fourth Dimension. Oscarovitch was becoming more and more fascinated as the light-winged minutes sped by, and he took but little pains to conceal the fact. Nitocris, of course, saw this, and simulated a delightful unconsciousness. The Professor was, for the time being, completely mystified. He knew that his daughter hated the Prince with a thorough cordiality, and yet he had never seen her make herself so entirely charming to any man, not even excepting Merrill himself, as she was to this man, her enemy of the Ages. He could have solved the problem instantly by crossing the Border, but then the sudden vanishing of a famous scientist from the midst of the brilliant company on the balcony would have set all the newspapers in Europe chattering, with consequences which would have been the reverse of pleasant both to his daughter and himself.

However, he had not long to wait, for Nitocris soon rose, saying that she must go to Jenny, her maid, to see about packing arrangements for to-morrow; and the Prince, after another cigarette and liqueur, took his leave and went on board the yacht to give orders for her to be put into her best trim, and then to have a luxurious half-hour with the Horus Stone, and indulge in fond imaginings as to how it would look hanging from a chain of diamonds on the white breast of Miss Nitocris.

When the Professor went to his own sitting-room he found his daughter waiting to say good-night.

"Niti," he said, as he closed the door, "I don't want to seem inquisitive, but, frankly, I was astounded at the gracious way in which you treated that scoundrel Oscarovitch."

"Dad," she replied, with apparent irrelevance, "do you believe in the forgiveness of sins?"

"Of course not! How could any one who holds the Doctrine do that? We know that every moral debit must be worked off and turned into a credit by the sinner, however many lives of suffering it takes to do it. Why do you ask?"

"So that you might answer as you have done!" she said, with a little laugh. "Now this Oscarovitch has sinned grievously, not only in this life but in many others, and I am going to see that he works off at least some of his debit as you put it somewhat commercially. He loved me in the old days in Memphis, and he loves me still in the same brutal, animal way. I know that if he cannot get me by fair means he will try to take me by force—and I am going to let him do it."

"Niti!"

"Yes, he shall take me; he shall think he had got me safe away from you and Mark—and when he has got me he shall taste what the hot-and-strong sort of Christian preachers call the torments of the damned. No, I shall not kill him. He shall live till he prays to all his gods, if he has any, that he may die. He shall hunger without eating, thirst without drinking, lie down without sleeping, have wealth that he cannot spend, and palaces so hideously haunted that he dare not live in them, until, when men wish to illustrate the uttermost extreme of human misery, they shall point to Prince Oscarovitch. I, the Queen, have said it!"

Then, with a swift change of voice and manner, she laid her hands on her father's shoulders, kissed him, and murmured:

"Good-night, Dad—at least as far as this world is concerned."

CHAPTER XXI
WHAT HAPPENED AT TRELITZ

It was the 6th of June again.

Once more Prince Zastrow rode with Ulik von Kessner and Alexis Vollmar and the attendant huntsmen up the avenue of pines leading to the gate of the Castle of Trelitz, but now accompanied by two unseen Presences which belonged at once to their own world and also to another and wider one. Once more the great doors opened and they passed into the trophy-decked, skin-carpeted hall: and once more they were welcomed by the stately, silken-clad woman who came down the broad staircase to greet her lord and his guests. Emil von Zastrow, last and worthiest scion of his ancient line, the very *beau ideal* of youthful strength and manly dignity, ran half-way up the stairs to meet his lady and his love, and then the men went away to their rooms, while the Princess Hermia, true housewife as well as princess, betook herself to the pleasant task of making sure that all the preparations for dinner were complete.

The dinner was served in one of the smaller rooms, in the modern wing of the Castle, on an oval table. The Prince sat at one end faced by his beautiful consort. To his right sat his guest, Alexis Vollmar, and a tall, handsome, but somewhat hard-featured woman of about thirty, with the clear blue eyes and thick, yellow-gold hair which proclaimed her a daughter of the northern German lowlands. This was Hulda von Tyssen, the Princess's companion and lady-in-waiting. They were faced by a stout, powerfully-built man with a full beard and moustache *à la* Friedrich, Ulik von Kessner, High Chamberlain of Boravia. Captain Alexis Vollmar was a typical Russian officer of the younger school, tall, well-set-up, and good-looking after the Muscovite fashion. He had distinguished himself in the Far East, but just now he preferred the serene atmosphere of Boravia to the thunder-laden air of Holy Russia.

The talk was of hunting and war and politics and the chances of the Russian revolution, and on this latter subject it was perfectly unrestrained, for all knew that the Powers had made a secret compact by which they bound themselves, in the event of the fall of the Romanoff Dynasty and the Arch-Ducal oligarchy—which all Europe would be very glad to see the last of—to support Prince Zastrow as elective candidate for the vacant throne.

The Revolutionary leaders had been sounded on the subject, and were found strongly in favour of the scheme. It meant a return to the ancient principle of elected monarchy, and Prince Zastrow, though now a German ruling prince, represented the union of two of the oldest and noblest families in Russia and Poland. Moreover, he had pledged himself to a Constitution which, without going to Radical or Socialistic extremes, embodied all that the moderate and responsible adherents of the Revolutionary cause desired or considered suitable for the people in their present stage of political development—which, of course, meant everything that Oscar Oscarovitch did not want.

After dinner they went out through the long French windows on to a verandah which overlooked a vast sea of forest, lying dark and seemingly limitless under the fading daylight and the radiance of the brightening moon. Since their marriage day the Prince had made it a bargain that whenever they dined *en famille*, his wife should prepare his coffee with her own hands. She even roasted the berries and ground them herself, and, as many a time before, she did it to-night in the seclusion of the little room set apart for that and similar purposes. She was alone in the physical sense, for the two watching Presences were invisible to her, and so, for all she knew, no one saw her measure twenty drops of a colourless fluid from a little blue bottle into the coronetted cup of almost transparent porcelain which had been one of her wedding presents to her husband.

After a couple of cups of coffee and half a dozen half-smoked cigarettes, the Prince stretched his long legs out, struggled with a yawn, and said in a sleepy voice:

"My Princess, you must ask our guests to excuse me. I am tired after the long day in the sun; and so, if I may, I will go to bed."

He rose, and the rest rose at the same moment. He bowed his good-night, and the two saluted. The Princess followed him into the dining-room.

The unseen watchers stood by the end of the great heavily-hung bed, in the midst of which lay Prince Zastrow, seemingly sinking into the slumber of death. Von Kessner leaned over and raised an eyelid, and said to the Princess, who was standing on the other side, the single word: "Unconscious." She bent forward for a moment as though she were bidding a silent farewell to the man to whom she had pledged her maiden troth, then straightened up and walked like some beautiful simulacrum of a woman towards the door which Vollmar held open for her....

The earth-hours passed, and the two men kept their watch by the bed, conversing now and then in whispers between long intervals of anxious silence, until three strokes sounded from the bell of the Castle clock. The whole household, save one fair woman, who, in softly-slippered feet, was pacing the floor of her bedroom, was fast asleep, and the days of sentries were far past. Von Kessner gently lifted one of the arms lying on the coverlet of the bed and let it fall. It dropped as the arm of a man who had just died might have done. Again he raised an eyelid, this time with some difficulty. The eyeball beneath was fixed and glassy as that of a corpse. He nodded across the bed to the Russian, and together they turned the bedclothes down to the foot. Then from under the bed he pulled out a bundle of grey skins which he spread on the floor beside the bed. It was a sleeping bag such as hunters use in winter on the snow-swept plains and forests of Northern Europe. Vollmar turned the head-flap back. Then they lifted the body of the Prince from the bed, slid it into the sack, and buttoned the flap down over the face.

"That Egyptian's drug has worked well," whispered Von Kessner.

Vollmar nodded, and whispered back:

"I wish I had a handful of it. But it is time. He will be ready for us now."

Even as he spoke the locked door opened, as it were of its own accord, and Phadrig stood in the room dressed in the livery of the Prince's coachman. Von Kessner and Vollmar turned grey as he bowed, and whispered:

"The doors are open, Excellencies, and all is ready!"

Then the three lifted the shapeless bag and carried it with noiseless tread down to the hall and out through the half-open doors to where a carriage drawn by four black horses stood waiting. Though there was no one in charge of them, they stood as still as though carved out of blocks of black marble until the body of the Prince had been laid in the carriage and Von Kessner and Vollmar had taken their places beside it. Then Phadrig mounted the box, shook the reins, and the rubber-shod horses moved silently away at a trot, which, as soon as the main road was reached, became a gallop only a little less silent than the trot.

The carriage turned aside from the road, and ran down a broad forest lane till it stopped by the shore of a little sandy inlet. The bow of a long black boat was resting on the sand, and six closely-blindfolded men were sitting on the thwarts with oars out. Another stood on the beach with the painter in his hands. The body of the Prince was carried from the carriage to the boat, and laid in the stern sheets. Von Kessner and Vollmar remained on board, and Phadrig went back to the carriage. At a short word of command the oarsman backed hard, and the boat slid off the sand into the smooth water of the little cove. Then she shot away and melted into the light haze which hung over the outside sea.

The boat stopped under the shadow of the long, low-lying black hull of a four-funnelled destroyer. A rope dropped from the deck and was made fast by Vollmar in the bow. The blindfolded crew were helped up the ladder which hung over the side and taken below forward. Then came a sharp order: "All hands below"; and when the deck was deserted, Von Kessner and Vollmar went up the ladder and were met on deck by Oscar Oscarovitch in civilian dress. There was another man beside him in the uniform of a lieutenant. He slacked off the tackle falls of the davits under which the boat had brought up, dropped down the ladder and hooked them on. When he got back to the deck the four men hauled first on one tackle and then on the other, till the boat was up flush with the deck. The falls were belayed, and Oscarovitch got into the boat and opened the flap of the sleeping-sack. He touched the spring of an electric pocket-lamp and looked upon the calm, cold features of his rival. Then he buttoned down the flap again and returned to the deck. The four went down into the cabin: glasses were filled with champagne, and as Oscarovitch raised his to his lips, he said:

"Count and Captain Vollmar, I am satisfied. Let us drink to the New Empire of the Russias and the sceptre of Ivan the Terrible!"

"And his illustrious successor!" added Von Kessner.

Within half an hour a small boat was lowered; the Chamberlain and Vollmar got into it and rowed away toward the cove. The Russian officer went on to the little bridge, signalled "full speed ahead" to the engine-room, and then took the wheel. The screws ground the water astern into foam, the black shape leapt forward and sped away eastward into the glimmering dawn with its silent passenger lying in the swinging boat, and the unseen watchers standing by the helmsman....

More earth-hours passed. The sun rose upon a lonely sea. The destroyer stopped, and a white speck on the eastward horizon rapidly grew into the white shape of a large yacht flying through the water at a tremendous speed. In a few minutes she was almost alongside. She swung round in a sharp curve, slowed down and dropped a boat. Oscarovitch and the lieutenant lowered the destroyer's boat till it touched the water. The other came alongside, and the body of Prince Zastrow was transferred to it, and Oscarovitch followed it. Four men from the yacht's boat jumped on board the destroyer and hauled hers up. The other was backed to the ladder and they came on board. A silent salute passed between Oscarovitch and the lieutenant, and a few minutes later the yacht's boat was hoisted to the davits, and the white shape was growing smaller and dimmer amidst the light haze that lay on the water shimmering under the slanting rays of the rising sun.

Morning grew into noon, noon faded into evening, and evening darkened into night. The yacht ran into a wide-opening gulf between two forest-clad points, on the southern of which twinkled the lights of a large town. These were soon left behind by the flying yacht, and as a vast sea of fleecy cloud drifted up from the north-east and spread its veil across the path of the half moon, a little cluster of lights gleamed out on the port bow. Her bowsprit swerved to the left till it pointed directly to them. Presently she slowed down and ran into a little land-locked bay surrounded with dense pine woods which came down almost to the water's edge, swung round and slowed up alongside a wooden jetty. From this a broad road, cut straight through the forest, sloped steeply up to a plateau on which stood a gaunt, grey, turreted castle, the very picture of the sea-robbers' home that it had been in the days of Oscarovitch's not very remote ancestors. Up this road and into the outer gate across the lowered drawbridge the sleeping-sack and the insensible man within were borne. Through the keep-yard it was taken into the Castle and up to a large room in the eastern turret, comfortably furnished, and containing a bed almost as luxurious as that in which Prince Zastrow had lain down to sleep the evening before. Oscarovitch preceded the men who carried him, and was met at the door by a grey-haired, keen-eyed man, who bowed before him, and said in a low tone:

"May I presume to ask if this is my charge, Highness?"

"It is, Doctor Hugo; and I give him into your hands with every confidence that you will restore your patient to health as quickly as any man in Europe could do. I must leave immediately, and so I trust everything to you. All care must be taken of him. He must want for nothing that you can give him—except liberty."

Oscarovitch returned the doctor's assenting bow and left the room. In half an hour the yacht was flying at full speed over the smooth waters of the Baltic, heading a little to the south of West.

CHAPTER XXII
A TRIP ON THE SOUND

"Good morning, Dad," said Nitocris, as she entered the sitting-room about half an hour before breakfast the next morning. "What is your opinion of the European situation now?"

"Good morning, Niti; what is yours?" asked her father, looking at her with grave eyes and smiling lips.

"As it was yesterday, only rather more so. In his present incarnation, Prince Oscar Oscarovitch is, I should think, about as black-hearted a scoundrel as ever polluted the air that honest people breathe."

"I entirely agree with you. And now, believing that, do you still propose to trust yourself to his tender mercies on board his own yacht, surrounded, as you will be, by men who, no doubt, are his absolute slaves?"

"*I* trust myself to his tender mercies, Dad?" she replied, drawing herself up and throwing her head back a little; "you seem to have got hold of the thing by the wrong end, as Brenda would say. That is only what it will look like. The reality will be that he will blindly trust himself to *my* mercies—and I can assure you that he will find them anything but tender. No, dear, we shall accept His Highness's invitation to lunch, and then his offer of the hospitality of the yacht for the trip, which, by the way, I fancy will be more to the eastward than to the northward— —"

"You mean, I suppose, Trelitz and Viborg?"

"Not Trelitz, I think, but Viborg almost certainly. That will be the end of the abduction as far as I can see from our present plane of existence."

"Really, Niti—well, well. Of course, I know that you will be perfectly safe: but what would our good friends on this plane, as you put it, the Van Huysmans, for instance, think if they could hear you talking so calmly to your own father about getting yourself abducted by a man whom you justly think to be one of the most unscrupulous scoundrels on earth! And, by the way, what is to become of me in the carrying out of this little scheme of yours? I hope you don't expect me to connive at the abduction of my own daughter. I have a certain amount of reputation to lose, you know."

"Oh, if His Highness is the clever villain that we know him to be, I think we may safely trust him to arrange for your temporary disappearance from the scene. And whatever he does it will be easy for you to play the part of the passive victim for the time being. He can't injure or kill you, for if it came to extremities you have the means of giving his people such a fright as would probably drive them out of their senses, just as I could if their master got troublesome. Really, from a certain point of view, the adventure will have a decidedly humorous aspect."

"With a very considerable leaven of tragedy."

"Yes, the tragedy will be a logical sequence of the comedy—and, as I said last night, it will be tragedy. And now suppose we go to breakfast. I have been up nearly two hours helping Jenny with the packing, and this lovely air has given me a raging appetite. There's a little more to do yet, and we shall have His Highness here before long to ask for our decision and take us off to the yacht."

Here she was quite right, for she had hardly left her father to his after-breakfast pipe and gone upstairs to help her maid, than Oscarovitch came into the smoking-room.

"Good morning, Professor Marmion! I need not ask you if you have had a good night. You look the very picture of a man who has slept the sleep of the just. And Miss Marmion?"

"Thanks, Your Highness, I think we have both managed to spend the night to good purpose. The air here is glorious just now. I always think that sound, dreamless sleep is the best sign that a place is doing you good."

"Oh, undoubtedly, though for some reason or other I did not sleep very well last night. Something had disagreed with me, I suppose. I seemed to have a sense of being pursued to the uttermost ends of the earth and back again by some relentless foe who simply would not allow me to take a moment's rest. But I didn't come to talk about the stuff that dreams are made of. I came to ask whether my cruise is to be a lonely one, or whether I am to have the very great pleasure of your company."

Franklin Marmion, for perhaps the first time in his life, felt distinctly murderous towards a fellow-creature as he looked at this splendid specimen of physical humanity, knowing so well the real man who was hiding behind that fascinating exterior; but he managed to answer pleasantly enough:

"We have talked the matter over, Prince, and we have come to the conclusion that your very kind invitation is really too good to be refused. We know that we are incurring a debt that we shall not be able to pay, but we are trusting to your generosity to let us off."

"On the contrary, my dear Professor," said Oscarovitch, without the slightest attempt to conceal the pleasure that the acceptation gave him, "it is yourself and Miss Marmion who have made me your debtor. In fact, if you had not found yourselves able to come, I should have run the *Grashna* back to Cowes, gone up to London, plunged into a maelström of dissipation, and probably ended by losing a great deal of money at Ascot and Goodwood. Ah, Miss Marmion, good morning! How well the air of Copenhagen seems to agree with you! The Professor has just gladdened my soul by telling me that you have decided to take pity on my loneliness."

"Good morning, Prince!" she replied, putting her hand for a moment in the one he held out. "Yes, we are coming, if you will have us. In fact, I have just finished packing."

"Ah, excellent! Well now, since that is happily arranged, it would be a pity to waste any of this lovely morning. The Sound is like a streak of blue sky fallen from heaven. My gig is down at the jetty, and I have a couple of my men here who will convoy your baggage down. If it is packed, as you say, you need not trouble about it. You will find everything safe on board."

"Thank you, Prince," said the Professor. "Then I will go and settle up at the office while Niti puts her hat on. I will have the things sent down, and we may as well walk to the jetty. It will do me good after that big breakfast. Jenny had better get into a cab and go down with the luggage."

When they reached the promenade along the Sound shore Oscarovitch pointed to a beautifully-shaped, three-masted, two-funnelled white yacht lying about five hundred yards out, and said:

"That is the *Grashna*, Miss Marmion. I hope you like the look of her."

"She is beautiful!" exclaimed Nitocris, recognising at once the vessel which had met the Russian destroyer on the early morning of the 7th. "She almost looks as if she could fly."

"So she can in a sense," laughed the Prince. "Come now, here is the gig. We will get on board, and you shall see her go through her paces."

Neither she nor her father were strangers to yachts, but when they mounted the bridge of the *Grashna* and looked over her from stem to stern, they had to admit that they had never seen anything quite so daintily splendid. They had chosen their rooms, and Jenny was below unpacking. Although, of course, he had a captain on board, the Prince often sailed the yacht himself when he had guests on board. He had a genuine love for the beautiful craft, and he took an almost boyish delight in showing what she could do. She was a twelve-hundred-ton, triple-screw, turbine-driven boat, and, thanks to the space-economy of the new system, her builders had been able to stow away fifteen thousand horse-power in her engine-room, and this when fully developed gave a speed in smooth water of thirty-five knots or a little over forty statute miles an hour.

The anchor was up almost as soon as they got on to the bridge, and Oscarovitch moved the pointer of the telegraph to "Ahead slow." The quartermaster in the oval wheel-house behind him moved the little wheel a few spokes to starboard, her mellow whistle tooted, and she glided in an outward curve through the other yachts and shipping, and gained the open water.

"Now," he said, turning to Nitocris, "we can begin to move. It is roughly thirty English miles to Elsinore. If you have never done any fast travelling at sea and would like to do some now, I can get you there in about three-quarters of an hour."

"What!" exclaimed the Professor, "thirty miles in forty-five minutes by sea! That is over forty miles an hour. A wonderful speed."

"Yes," he replied, almost tenderly; "but my beautiful *Grashna* is a wonderful craft—at least, I think you will say so when you see what she can do. Now, if you will take advice, you and Miss Marmion will go into shelter, for it will begin to blow soon."

Behind the wheel-house was an observation room, as it would be called in the States, running nearly the whole length of the bridge, and fronted with thick plate glass. They went in, and Oscarovitch turned the pointer to half-speed. There was no increase in vibration, but the shore began to slip away behind them faster and faster, and the northern suburbs of Copenhagen rose ahead and fell astern as though they were part of a swiftly moving panorama. Then the pointer went down to full speed, and the Prince, after a word to the quartermaster, joined them in the bridge-house and closed the door.

"You will need all your eyes to see much of the shore now," he said; "I have given her her wings."

Nitocris felt a shudder in the carpeted floor. Looking ahead she saw the bow lift slightly. Then a smooth, green swathe of water curled up on either side. She looked aft, and saw a broad torrent of froth, foaming like a furious, rapid stream away from the stern. The houses and trees on the shore seemed to run into each other, and slide out of sight almost before the eye could rest upon them. The water alongside was merely a blue-green blur. Nitocris involuntarily held her breath as though she had been out on deck.

"It is wonderful, Prince!" she said, almost in a whisper. "That alleged express from Hamburg was nothing to this: and yet how steadily she moves in spite of the speed. I should have thought that it would have nearly shaken us to jelly."

"That is the turbines, dear," said her father, who was already wondering whether Oscarovitch was doing this just to show how hopeless any pursuit of such a vessel would be. "They are a marvellous means of applying steam power. Lieutenant Parsons is robbing the sea of one, at least, of its worst terrors."

"Yes," added the Prince, "we are travelling a little over forty miles an hour; and if you got that speed out of reciprocating engines you would scarcely be able to lie on the deck without holding on to something, yet here we are as comfortable as though we were standing in a drawing-room."

"You have given us a new experience to begin with," said Nitocris, thinking how nice it would be to take her wedding trip with Merrill in such a craft as this. "Why, look at the two shores coming together, Dad!"

"No, excuse me," said Oscarovitch, "we are only about half-way to the Gate of the Baltic yet. That land on the right is the island of Hvreen. When we have passed that you will soon see the heights of Elsinore and Helsingborg rising ahead. There are only about two and a half miles between Denmark and Sweden there."

"Oh yes, of course. I am forgetting my geography," laughed Nitocris, as the low, wooded patch of land came rushing towards them as though it were adrift on a fast-flowing stream. "Goodness, what a speed!"

"A very wonderful craft, Prince," added the Professor, as the island drifted past; "she quite inclines me towards a breach of the tenth commandment. Now that you have given us this taste of the delights of speed, I think that if I were a millionaire, I should try to build one to beat her."

"Exactly," laughed Oscarovitch. "It is marvellous this fascination of speed. Your poet, Henley, touched the pulse of the times when he wrote those splendid lines of his. But surely, Professor, *you* would not have very much difficulty in leaving all far behind. A man to whom mathematical impossibilities are as easy as an addition sum ought to be able to realise the dream of the ages and solve the problem of aerial navigation."

He looked him straight in the eyes as he said this. He fully believed in the possibility of human flight, given the transcendent genius who could work out the equation of weight and power. Perhaps that genius might be with him now in the bridge-house. His vivid imagination was already picturing the lovely girl at his side crowned Empress of the Russias and the East, and himself in command of an aerial navy, beneath whose assault the armies and navies and fortresses of the rest of the world would be as so many toys to play with and destroy.

"If I could do that, and I do not think it would be so very difficult after all," said Franklin Marmion, returning his glance, "I would not do it. It would put too much power in the hands of a few men, and we have enough of that already. The owner of a fleet of aerial warships would be above all human law. He could terrorise the earth, and make mankind his slaves. Life would become unendurable under such conditions. Commercialism, which only means slavery plus the liberty to starve, is bad enough, but it is at least possible. The other would be impossible. There is no man quite honest enough to be trusted with such a power as that. I have worked the thing out, and it is perfectly feasible, but I burnt my designs and calculations."

"What!" exclaimed Oscarovitch, flushing in spite of his effort to keep the blood back from his face. "You have solved the problem, and won't make use of the greatest invention of all the ages! Surely, Professor, that is a little quixotic, is it not?"

"Who am I that I should bring a curse upon humanity, Prince?" he answered gravely. "Do you not kill each other fast enough now? No, the world is not fit for such a development yet. My results will remain my own until Tom Hood's ideal of good government has been realised."

"And what was that, Dad?" asked Nitocris, who had a double reason for being interested in the conversation. "If I ever knew it, I have forgotten it."

"Despotism, Niti—and an angel from heaven for the despot," he replied, with another look into the Prince's eyes which brought him to the conclusion that the sooner his presence on board the *Grashna* was dispensed with the better for his plans. There was a sense of quiet mastery in Franklin Marmion's manner which made him uneasy.

"Ah! there is the famous fortress, is it not? the home of Hamlet and Ophelia and the Ghost!" she exclaimed, pointing ahead to where a grey-blue mass was rising out of the water. "Do you believe in ghosts, Prince?" she added suddenly, flashing a glance at him which seemed to pierce his brain like a ray of unearthly light.

"Ghosts? No, Miss Marmion. I'm afraid I am too hopelessly materialistic for that. I never saw or heard of an authentic ghost, and I do not propose to believe until I see."

"We have a ghost at 'The Wilderness,'—the wraith of a poor young lady who killed herself after some royal blackguard had abused his own hospitality. She often comes to visit me in my study," said the Professor, as though he were relating the most ordinary occurrence.

"Ah," smiled the Prince, "that is very interesting: but, of course, it would be in the power of a man like yourself to have experiences which are denied to ordinary mortals. Still, granted all that, I confess that I have often wondered whether or not I should be frightened if I really did see a ghost."

"Yes, I wonder?" murmured Nitocris, with a great deal more meaning than he had any idea of just then.

All three felt that the conversation was getting a little difficult, and they were not sorry when the rapid rising of the rock of Elsinore made it necessary for Oscarovitch to go out to the engine telegraph.

"His Highness doesn't believe in ghosts now," whispered Nitocris to her father, when the door shut behind him, "but I think he will before very long. I wonder what he is really going to do? I've half a mind to— —"

"No, no, Niti," he said quickly; "keep this side of the Border till you really have to cross it. What on earth, literally, would happen if he came back and found me standing here alone?"

"Oh, of course I didn't mean it," she smiled. "It would be very poor sport to spoil both the comedy and the tragedy before the curtain goes up. I wonder if the drama will begin to-night? I shouldn't be surprised."

"Nor I," said the Professor, a trifle grimly. "I didn't at all like his looks when I was talking about the flying machine. The brute looked as if he were quite capable of locking me up and starving or torturing me until I gave him the secret. My word, I should like to see him try! I'd have him grovelling at my feet in five minutes."

The door opened and Oscarovitch came in. He took off the cap which had been pulled tight over his eyes, and said:

"Well, we have arrived! Almost exactly forty-five minutes. There is Elsinore, there is Kronborg, King Frederick's sixteenth-century castle, and there is Marienlyst, which is to Copenhagen what Brighton is to London, only, I must say, in a much more refined sense. Now what is your pleasure,

Miss Marmion? We have still nearly two hours before lunch, so, if you would like an hour's stroll ashore, the gig will be ready in a couple of minutes."

"Thank you, Prince," she said with a rewarding smile. "Dad, what do you think? It all looks very beautiful under this sun and sky."

"Which, of course, means that you want to go ashore, Niti," said her father. "For my own part, I certainly should like a little walk on new ground. I have never been here before."

"Then, of course we will go," said Oscarovitch, opening the door and going to the telegraph.

The yacht came to a standstill in a few minutes, and the gig was waiting at the foot of the gangway ladder. They spent a very pleasant hour ashore, and what they saw, you may read of in your Murray and Baedeker, wherefore there is no need to set it down here. When they came aboard again, lunch was almost ready, and the steward presented his master and the Professor with quite exceptional cocktails in the smoking-room. Then they went and had a wash, and the mellow gong sounded.

I am not very fond of those descriptions in stories which read like extracts from an upholsterer's price-list, nor yet those accounts of meals that, after all, are only menus writ large, so it may suffice to say that the saloon of the *Grashna* was an arrangement of sandal-wood panels, framed in thin silver filigree, and hung with exquisite little masterpieces in water-colour, and black and white, and crayon, mostly sea-scapes, with here and there a beautiful head with living eyes which followed you everywhere; that the rich yellow of the panels was enhanced by *portières* and curtains of deep golden-bronze silk, and that the domed ceiling was of pale, sky-blue enamel spangled with the constellations of the northern heavens, which at night lit up the whole saloon with a soft electric radiance. As for the lunch, it was as nearly perfect as the best-paid chef afloat could make it, after his master had asked him as a personal favour to do so.

They ran back quietly to Copenhagen at twenty knots, and Oscarovitch and the Professor went ashore to send off a few telegrams, leaving Nitocris, for her own reasons, to make herself at home on the yacht. They returned in time to dress for dinner and enjoy a stroll on the broad upper deck, and watch the sunset over the town and the quickly-increasing sparkle of the myriad lights on shore and sea. When they came up after dinner, these lights were only represented by a luminous haze glimmering under the stars to the northward. The *Grashna* was heading nearly due south at an easy speed towards the Baltic Islands.

Something told both Nitocris and her father that the decisive hour would come soon, and they were both prepared for its advent.

CHAPTER XXIII
THE DISAPPEARANCE OF THE PROFESSOR

The Prince and the Professor sat up in the smoking-room for a considerable time after Nitocris had retired. Oscarovitch was doing his utmost to persuade his guest to revoke his decision as to the creation of the aerial warships. Franklin Marmion's simple announcement, which he never thought for a moment of disbelieving, had filled his mind with new ideas, which were rapidly taking the shape of gorgeous dreams of an empire such as mortal man had never ruled over before. All his present designs faded away into mere trivialities in comparison with this splendid conception. He pictured Nitocris, as his consort, Empress of the air, and himself Lord of earth and sea and sky. But all his subtle arguments, all his delicately-put suggestions, and his skilfully framed promises failed to produce the slightest effect upon the genially inflexible man, who quietly turned them all aside, as a grown man might deal with the arguments of a boy.

The thought that this man who was lying back in his deep-seated armchair, holding a cigar in a white, delicately-shaped hand which was strong enough to shake the world to its foundations, should possess such a tremendous power and yet refuse to use it, as quietly as he might have declined an invitation to dinner, exasperated him almost beyond the bounds of patience. If he would only join forces with him what glories might they not achieve, what splendours of power and possession might not be theirs! Here was universal empire, in one sense, only a couple of yards away from him! In another it was more distant than the suns which flame in Space beyond the Milky Way. It was maddening, but it was true, and he knew the man well enough now to feel absolutely assured that no extremity of mental or physical torment would wring the priceless secret from him.

Well, if it had to be, it must be. If he could not learn the secret, at least no one else should. Before morning it would be buried for ever under the waters of the Baltic, and he would revenge himself on the daughter for that which the father refused to do. If Franklin Marmion would not give him the sceptre of the World-Empire, then Nitocris should be his wife and Empress if she would, and if not, his slave and plaything, as he had sworn to Phadrig the Egyptian. The fortress-castle of Oscarburg, on the lonely wooded shore of Viborg Bay, had kept many a secret safely before now, and it would keep this one. Every retainer in the Castle, every man, woman, and child on the estates for leagues around, was his, body and soul, as their fathers before them had been the blind, unquestioning serfs of his fathers. There his word was law, and his will was fate. There was no "liberty" within his domains, since no man wanted it, or would have understood it had it been given to him.

When their argument was over they parted, apparently the best of friends. Franklin Marmion went to bed calmly curious as to what was going to happen, and Oscarovitch paid a visit to his captain.

A little after three that morning he opened the door of the Professor's state-room very gently and looked in. The room was dark, and he listened. A soft, just audible sound of breathing came from the bed. It was the breathing of a man fast asleep. He pressed the spring of his electric lamp, and turned the thin ray on to the water-bottle in the rack over the wash-stand. It was half-empty, and a glass stood on the table in the middle of the room. Then the ray fell on the face of the sleeping man. It was as Prince Zastrow's face had been the last night he went to sleep in the Castle of Trelitz—rather the face of a corpse than that of a living man. His captain stood behind him, and he turned and whispered:

"He is ready. Are the men below?"

"All, Highness, save Grovno at the wheel and Hartog on the look-out. They will see nothing, as they did before," came the whispered reply.

"Very well, then. You and I can manage this between us. You have the line?"

The captain nodded, and they went into the room, softly closing the door. In a few minutes they came out again, carrying between them a long bundle of blankets lashed from end to end with thin line. They took it aft along the alloway and out on to the lower deck by the stern. Two iron doors of a port used for coaling stood open on the starboard side. On the deck lay a couple of pigs of iron lashed together. These the captain made fast to one end of the bundle and lifted them towards the port. Oscarovitch took hold of the other end. They lifted it. The weights dropped outside the port, and the bundle followed them. The captain started up, clasped his hands to his forehead, and said in a gasping whisper:

"Holy God, Highness, what have we done?"

"What do you mean, Derevskin? You have obeyed my orders; that is all. Is it not enough for you?"

"Yes, Highness—but who or what was that man? Was he really a man?"

"Are you mad, Derevskin?"

"No, Highness, I hope not: but did you hear—or, rather, did you not hear?"

"What, you fool?"

"He—it—the body—it made no splash when it touched the water!"

The stammered words struck Oscarovitch like so many puffs of frozen air. No, the body of Franklin Marmion *had* made no splash. It had vanished through the port into silence. That was all. He beat back his own terror with the exertion of all his will-power, and said in a sneering whisper:

"Derevskin, you are either mad or drunk; but I will forgive you this time because you have obeyed. Go to bed, and don't forget to be either sober or sane when I come on deck."

The captain bowed his head, and went forward with shambling steps and shaking limbs. Oscarovitch closed the port with hands which all his force could not keep steady, and betook himself to bed, to lie awake for the rest of the short summer night wondering vainly what really had happened.

He had had his bath and dressed soon after six, and went on deck. The captain was on the bridge, and he joined him.

"Good morning, Derevskin!"

"I have the honour to wish Your Highness good morning!"

"Nothing happened during the night worth reporting, I suppose?"

"No, Highness, nothing."

"Very good: but I have slept badly, and you look as if you had been on the bridge all night. Perhaps it is necessary among all these islands, and I am pleased that you are so watchful, especially as I have guests on board. Come down to your room now and send your steward for a bottle. It will do neither of us any harm."

There was a somewhat lengthy conversation over this early breakfast of champagne and biscuits after the door had been closed and locked, and when it was finished, Oscarovitch and his captain understood each other as completely as was necessary.

An hour later he saw Nitocris walking about the upper deck looking pale and anxious. He went to her and said in a tone which intentionally betrayed his own nervousness:

"Good morning, Miss Marmion! Have you seen anything of the Professor?"

"No, Prince, I have not. I went to his room just now and knocked. There was no reply and I opened the door. The room was empty, but he had evidently been to bed. Is he not on deck?"

"No, Miss Marmion, he is not. He said last night that he would like his bath about six, and the steward I sent to valet him went to his room and found it as you say. I have had the ship searched high and low, and from stem to stern, and there is no sign of him. I have had every one questioned, and no one has seen anything of him since last night."

"Oh, my poor, poor Dad, I have lost him! Yes, I suppose it must have been that. He has walked overboard."

"Walked overboard, Miss Marmion?"

"Yes, yes, it must be that. Prince Oscarovitch, my father, like most very clever men, had one dangerous failing. He walked in his sleep and did things unconsciously. That was why he told you about the ghost at 'The Wilderness' just as though he really had seen it. Yes, he must have got up in the night and come on deck, and walked overboard, and so I have lost the best friend I ever had, or shall have. You must excuse me, Prince. I must go to my room. The very sunlight seems horrible now. Jenny will look after me. Good morning!"

Her face was white and her eyes were staring at nothing. She spoke with a horrible, stony calm which, crime-hardened as he was, sent a thrilling shiver through his nerves. A spasm of remorse shook him; then his self-control came back, and he offered her his arm in silence. He led her down to the saloon, and gave her into Jenny's charge. Then he went on deck again, lit a cigar, and proceeded to congratulate himself on the great good fortune which had, from his point of view at least, so happily explained away the disappearance of Franklin Marmion.

CHAPTER XXIV
THE LUST THAT WAS—AND IS

Nitocris kept her room until nearly seven the following evening. Oscarovitch made frequent enquiries of Jenny as to her condition, and always received the same reply. Her mistress was in a semi-unconscious state, and she could only rouse her every now and then to take a little nourishment. Unfortunately there was no doctor on board. He had had news in Copenhagen that his mother was lying very ill at Hamburg, and, as the cruise was then intended to be only a very short one, he had been given leave to go to her.

The Prince wished to go back to Copenhagen, but this Nitocris absolutely refused. She had determined to fight her sorrow alone, and when she had conquered it, she would go back to England and her friends—which was exactly what Oscarovitch had determined she should not do. She was absolutely at his mercy now. He would be something worse than a fool to let such a golden opportunity go by—and so the *Grashna's* bowsprit was kept pointing eastward, and the leagues between her and Oscarburg were being flung behind her as fast as the whirling screws could devour them.

The only question that he had to ask himself was: How? and to that an easy answer at once suggested itself: The Horus Stone.

When he went down to what he expected would be a lonely dinner, he was more than agreeably surprised to find Nitocris dressed in a black evening costume, which was the nearest approach to mourning that her available wardrobe made possible, already in the saloon.

He bowed to her with a gesture of reverence, which meant far more than mere formal politeness, and said in a low tone:

"Miss Marmion, I need not say how pleased I am to find that you are able to leave your room. May I hope that you will be able to dine?"

"Yes, Prince," she replied, in the same cold, mechanical voice in which she had answered the tidings of her father's death. "The worst is over now, I hope. Some time and some way we must all leave the world and, at least, there is the consolation that my father has left it perhaps a little better and a little wiser than he found it. That, I think is as much as the ordinary mortal may be permitted to hope for. We who hold the Doctrine do not sorrow for the dead: we only sorrow for ourselves who are left to wait until we may, perhaps, meet again."

"The Doctrine, Miss Marmion?" he asked, as he placed a chair for her at his right hand. "May I ask what the Doctrine is?"

"Of re-incarnation," she replied, sitting down and looking at him across the corner of the table.

"Really? I most sincerely wish that I could believe in it. Mr Amena, whom I took the great liberty of bringing to your garden-party, a man of very remarkable powers, as you saw, holds the Doctrine, as you call it, and he has been trying for months to convert me to it; but, as I said going to Elsinore, I'm afraid I am too hopelessly materialistic for any conversion to be possible in my case, at least as far as my present experiences have gone."

"As the belief so must be the faith," she said with a grave smile. "It is no more possible to have true faith when you do not really believe than it is to be hungry when you have not got an appetite. That is quite a material simile; but I think it is true."

"Absolutely true!" he replied, looking at her again with a note of interrogation in each eye. "But, really, these things are too deep for me, a mere human animal. And now, talking about appetite, here comes the soup."

The dinner à *deux* was just what he had intended it to be, simple and yet perfect in every detail. The subject of Franklin Marmion's departure from the world was, as if by mutual consent, dropped. Oscarovitch comforted such conscience as he had by trying to believe that what Nitocris had said about her belief in the Doctrine was to her really true. He also honestly believed that she had faced her great sorrow in solitude, and overcome it in the strength of that belief. Their conversation turned easily away to other topics, and by the time that coffee was brought in and he had obtained her permission to light a cigarette, his beautiful guest appeared to have left the recent past behind her, for the time being at least, and was almost as she had been during the run up to Elsinore.

Her manner was that of complete composure, and it is hardly necessary to say that this mastery of her emotion forced him to a degree of admiration, almost of worship, which the physical charm that appealed only to his animal senses could never have inspired. Here, truly, was the ideal Empress of the

Russias and the East sitting almost beside him. And now the psychological moment had come!

"Will you excuse me for a couple of minutes, Miss Marmion?" he asked, as he finished his coffee and rose from his chair. "Going back to what you were saying about re-incarnation: I have something in my room which I hope may interest you. I got it from my friend, the miracle-worker. He told me a long story about it that I don't want to trouble you with: but the thing in itself is quite worth seeing. At least, I never saw anything like it before."

"Then please let me see it," she replied, assenting with an inclination of her head. "If that is so it must be, as you say, well worth seeing."

He went to his room and came back with a large square morocco case in his hand. He gave it to her, and said:

"Do me the favour to open it, and tell me what you think of it."

She touched the spring and the cover flew up. She half-expected what she saw. There, lying in a nest of soft black velvet, encircled by a triple halo of whitely gleaming diamonds, was the Horus Stone. In an instant she travelled back through fifty centuries to the scene of the death-bridal of her other self, Nitocris the Queen, in the banqueting-hall of the Palace of Pepi. Then it had lain gleaming on her breast, and now she saw it again with the eyes of flesh, after nearly five thousand years. Now, too, she grasped in all the fullness of its evil meaning the reason why Oscarovitch had brought it to her in such an hour as this. With utter contempt in her soul and a smile on her lips, she leaned back in her chair and said in a voice which had a note of ecstasy in it:

"Oh, Prince, how lovely! What a glorious gem! The diamonds are, of course, splendid, but they are only a setting for the emerald. What a magnificent stone! Rich as you are, you are very fortunate to be the possessor of such a treasure—for treasure it surely must be."

"It is, as you say, a magnificent stone," he replied, looking steadily into her questioning eyes. "But if what Amena told me was true, it is something more than a unique gem. There is an inscription on it, some characters carved in the stone which are, as he said, the history of it, but to me they are as unintelligible as the Assyrian cuneiform would be. Possibly you may know something of them. If you do, here is a lens that will help your sight."

She took the glass from him and bent down over the gem. She read the sacred symbol of the Trinity as she had read it and known it ages before. But while she was gazing at it, she also read the intent of the man who had given it into her hands. She put the lens aside, and, laying her palms on her temples, she looked deep down into the luminous depths of the great emerald in a silence which Oscarovitch interpreted into such meaning as he was able to make for himself.

Minute after minute passed in silence, and still her eyes were fixed upon the Stone. Her face became like that of a beautiful masterpiece of Phidias: pure, cold, and true. A feeling of something like awe crept over him as he watched her, and he found himself asking whether, after all, Phadrig's story might have been true. But, true or not, there was the fascination which, as Phadrig had told him, had lured Isaac Josephus to his self-inflicted doom. Her eyes were chained to the gem: her face was no longer that of a living woman dominated by her own will. After all his disbelief, there *was* an enchantment in the Stone, for here, even she, Nitocris, had succumbed to it.

He sat and waited for a few minutes longer. If there is magic in the Stone, let it work, he thought; and so he sat and watched her until he saw that the fixed stare of her eyes and the rigidity of her now perfectly statuesque face convinced him that the magic of the Stone had, as Phadrig had told him, made him the possessor of it, absolute master of the man or woman who had gazed upon its fatal beauty.

Then he got up and, reaching over her shoulders, took up the diamond chain, glistening under the soft light of the starry dome of the saloon, shook it out into a flood of white radiance, lifted it above her head, and let it fall very gently round her neck. The Horus Stone, as though endowed with sentience, fell and rested where it had rested five thousand years before. As it touched her flesh Nitocris felt a tremor of indescribable emotion, not only of the body but of the soul, pass through her. She leaned back in her chair again, and whispered:

"Is it really mine now, Prince? But no! How could I take it from you—I who can give nothing in exchange for such a treasure? No, no, you must take it back. I am not worthy to wear it."

He laid his hands gently on her arms, and said in a soft, murmuring tone which sounded like the purring of a tiger-cat:

"Nitocris, if all the choicest gems in all the world could be put into a crucible and fused into one, all its splendour would still be unworthy to lie on that white breast of yours. Give me your love, Nitocris. I am hungering and thirsting for it. Come with me to Oscarburg, and you shall be crowned Princess—and after that Empress—Empress of the Russias and the East. I will give you a dominion such as the great Catherine never dared to dream of. Say yes, and in a month you shall be seated on her throne. It is only a little word, dearest, only a little word—will you not say it, and be my Princess, my Queen, my Empress?"

"I am tired now, Oscar," she said wearily, "so much has happened in so short a time. Yes, I will, if it is possible: but let me go now. No, you must not kiss me yet. Remember that Russian saying, 'Take thy thoughts to bed with thee, for the morning is wiser than the evening.' Good-night, Oscar, I am very tired. You shall have your answer in the morning. May I take this with me?"

"Yes," he replied, giving her his hand as she rose from her chair, and bowing over hers until his lips touched it. "Take it, unworthy as it is, as an earnest of the realisation of the happy dreams that will come to me to-night. Au revoir, pas adieu!"

"Auf viedersehn, mein Oscar!" she replied as she passed him, leaving the sensation of a gentle flutter of her hand in his. "We shall understand each other better still before long—I hope."

"It is my dearest wish. Good-night, Nitocris, and when the dawn comes may it find nothing but sunshine in that sweet soul of yours!"

Nitocris went to her room and found her maid waiting, white-faced and anxious. She was frightened and nearly worn out with caring for her mistress. She would have been very glad to have been back that very night at "The Wilderness," even if it had lost its master.

"Go to bed at once, Jenny; you look like a ghost, as you may well do after all the trouble I've given you. No, I don't really want you, and you want sleep rather badly. Go to bed, like a good girl. It will not be the first time that I have undressed myself."

And when Jenny had gone and she had locked the door, Nitocris stripped herself, save for the collar of diamonds and the pendant Horus Stone. She took a long veil of Indian muslin out of her dress-box and wound it round her after the fashion of old Egypt, leaving her left breast bare. Only the Ureaus Crown was wanting to make her, in the flesh, Nitocris the Queen: but here on her bosom flashed and flamed the Horus Stone—hers once again, as it had been in the far-off past, symbol of her sovereignty, and proof of her faith in the one true Doctrine.

She looked at the lovely reflection in the long mirror behind her dressing-table, and said to herself in a low, whispering laugh:

"This for you, Oscar Oscarovitch that is, Menkau-Ra who was! Yes, you may dream your pleasant dreams to-night; you may take me to your lonely castle in Viborg Bay; you may make me marry you, as you think I shall—and here is my wedding gift—mine again after all these ages—blessed be for ever the Holy Trinity, Osiris, Isis, and Horus. May the Most High Gods help and protect me!"

She raised the Sacred Stone to her lips as she spoke, turned off the light, and lay down in her bed to dream dreams of forgotten ages.

CHAPTER XXV
THE PASSING OF PHADRIG

In all London, or, indeed, in any capital of Europe, there were no more angrily puzzled men than Nicol Hendry and his colleague and subordinates. He was perfectly certain now that Phadrig Amena held the key to the conspiracy which had resulted in the disappearance of Prince Zastrow. Oscarovitch had vanished. He had been traced to Copenhagen, and then absolutely lost sight of. Three agents, all picked experts, had been put on to watch Phadrig and the Pentanas, as they were known to him, and within a fortnight they had all died. One had fallen down crossing the north side of Trafalgar Square: the verdict had been heart failure. Another threw himself into the river from the Tower Bridge; and the third, a woman who was one of the most skilful spies in the service of the International, had made his acquaintance and had dinner with him at the "Monico," and was found dead the next morning with an empty morphia syringe in her hand and a swollen puncture in her left arm.

Thus four more or less valuable lives had been lost, and not a shred of tangible evidence obtained against the Egyptian. Convinced as he was that this man was as responsible for their deaths as he had been for that of Josephus, neither he nor his colleagues could find the slightest grounds for applying for a warrant for his arrest, and meanwhile things were going from bad to worse in Russia. The Romanoff dynasty was tottering to its fall. The responsible leaders of the Revolution, angry and bewildered by the loss of the man whom they had practically chosen to rule over them, were distributing thousands of copies of an unsigned manifesto which could not have come from any one but "the new Skobeleff." What was left of the army and the navy was rallying to the nameless standard of the still unknown saviour of Russia. Von Kessner and Captain Vollmar had apparently ceased to exist, and the Princess Hermia was living with her lady-in-waiting in the strictest retirement in Dresden.

"It seems to me that things are at an utter deadlock," said Nicol Hendry to the Chief of the German section, who had come over to London to confer with him. "Four of our best agents have died in a fortnight, and the others are getting shy. Really, we can't blame them. This is not like fighting the ordinary sort of anarchist or regicide, who, after all, does content himself with physical means. This infernal scoundrel, as I must confess I was warned to begin with, is quite independent of the rules of the game. He kills people by their own hands, not his, and, literally, there seems no way of catching him."

"There must be a way, my dear Hendry," replied the German, who was the very incarnation of mechanical officialism. "You look at these things as consequences, I regard them only as rather extraordinary coincidences. If this is anything like what you seem to think it, it is supernatural, and I don't believe in that."

"There is a very easy way to convince yourself, my dear Von Hamner," replied Hendry, with a slight shrug of his shoulders. "Suppose you go and interview this modern Mephistopheles yourself?"

"Will you come with me if I do?" asked the German, with a straight stare through his spectacles.

"Certainly. In our profession it is necessary to take risks. The thing has gone far enough. Here we are in my room at New Scotland Yard, the centre and stronghold of the British police system, and there is this man or super-man, if you like, making no sign, doing nothing that will give us a hold upon him, and yet killing our agents as fast as we send them to find out what he is working at, and we know just as much to-day as we did three weeks ago. Now, what is your idea?"

"Just this: if the English law won't touch him, do as we do in Germany, take the law into your own hands. We know where the fellow is to be found down in that slum near the Borough Road. Send a few of your plain-clothes men there this afternoon, and we will follow in a cab. Bring your bracelets with you, and I shall take my revolver. We don't want any nonsense this time. If it goes on much longer we shall be the laughing-stock of the whole force from end to end of Europe, and that will not do us any good. Shall it be for this afternoon?"

"It will be better done now. He has worked mischief enough, and if we are going to do it we may as well bring the thing to a head at once, as they say in the States. Now I will give the instructions, and we will go to lunch. It may be the last that either of us will eat, you know."

"Poof!" exclaimed Von Hamner, who was feeling not a little nettled at this quiet challenge to test his personal courage. "You are the last man on earth that I should have suspected of superstition, my dear Hendry. But, there, give your orders, and we will go to lunch, and then about four o'clock we may make our call in Candler's Court."

While the two Chiefs of the International were talking, Phadrig was reading a cypher telegram, of which the meaning was this:

"Reval.—Professor fell overboard three days ago. Body not recovered. Horus Stone did its work. N. consents. I marry her at Oscarburg. Russia ready. Fool International for a few days and come to Viborg when you have done with them.

O."

"That is good news," said Phadrig, in a confidential whisper to himself; "for a man on the lower plane of existence the Prince is wonderfully clever. This is a master-stroke. If he really has the Queen in his power all the rest will be easy."

"There's two gentlemen to see you, Mr Amena." The door opened, and his landlady's dirty little daughter put her towsled head through the little space behind the doorpost. "They're down below; shall I send 'em up?"

"Certainly, Jane. Tell the gentlemen that I shall be pleased to see them."

The dirty face vanished as the door closed. Phadrig shut down the top of the big escritoire and locked it. Heavy treads sounded on the rickety stairs. There was a shuffle of feet on the little landing, a sharp knock at the door, and he said in a low tone:

"Come in, gentlemen. I have been expecting you."

The door opened and Nicol Hendry entered, followed by his German colleague. Practised as they were in all the arts of their profession, they looked about the mean, miserably appointed room with curious eyes. Phadrig, dressed in the same shabby semi-Oriental costume in which he had received Isaac Josephus, salaamed, and said:

"Gentlemen, although this is but a poor room to receive you in, I am pleased that you have come. You are officers of the International, if I am not mistaken."

Then his speech changed to German, and he went on:

"You, sir, are M. Nicol Hendry, and your friend is the Herr von Hamner, Chief of the Berlin Section. What can I do to serve you?"

It was anything but the greeting that they expected. They thought that they had tracked the real criminal to his last hiding-place. They had established the identity between Phadrig, the poor seller of curios, and Phadrig Amena, the worker of miracles, whom all the smart set in London was talking about; and here he was in this miserable, shabby room, dressed in clothes that no pawnbroker would advance a couple of shillings on, smiling and bowing before them as though they were lords of the earth, and he—the man who had sent three men and a woman to their deaths by, as it were, a mere word of command—a worm beneath their feet. Nicol Hendry managed to keep his self-possession, but Von Hamner was already sorry that he had come, and his face showed it.

"We have come to ask you, Mr Amena," said Hendry, thinking it best to come to the point at once, "why you found it necessary to kill those people. I needn't mention names. You know them as well as we do."

"I did not kill them, gentlemen. They killed themselves, according to the newspaper reports. And now, may I ask you why you found it necessary to set these spies of yours to watch my every movement night and day? What have I done to bring myself within the four corners of your English law?"

"Nothing, unfortunately, that we can get a warrant for," replied Hendry, trying not to look into his eyes, "and so we have taken the law into our own hands. Come, Mr Amena, the game is up. We know all about your share in the conspiracy to remove Prince Zastrow in order to make room for your patron Prince Oscarovitch. We have copies of his manifesto at Scotland Yard, and we know that you received a telegram in cypher from him to-day."

"Ah!" said Phadrig, in a tone whose smoothness was intensely aggravating, "that is very interesting. May I ask if you have translated the cypher?"

"No, damn you and your Prince!" burst in Von Hamner. "If we had done that we should know even more about you than we do now—and that ought to be enough to hang you."

He had spluttered the words out before Hendry had time to stop him. He expected a tragedy there and then, but it did not happen. Phadrig took the telegram out of his coat pocket, handed it to Von Hamner with a graceful bow, and said:

"Your information is quite correct, gentlemen. That is the telegram, and this is the meaning of it."

Then as they read the unintelligible jumble of words, he repeated the meaning of them as though they formed the most ordinary message, instead of a dispatch that might, as they well knew, shake Europe to its social and political foundations within the next week or so.

"Then this is another of your devilries, I suppose," snarled Von Hamner. "So you have killed the great Professor Marmion, the most gifted genius in the whole world, as you killed the others, to promote your infernal schemes; and you have helped that scoundrel Oscarovitch to abduct his daughter. Well, law or no law, this shall be the end of your doings. You will come with us as our prisoner, or you will not leave this room alive."

"Those are hard words, mein Herr," said Phadrig, still speaking in German. "I your prisoner! Why? What have I done to make this outrage on English law possible?"

"You will do better to come, Mr Amena," said Hendry, in his quiet official tone; "it will save a good deal of trouble both to you and us. It must be the same in the end, you know. We have got you, and we don't mean to let you do any more mischief. You have done quite enough already. Now, will you come quietly, or shall we take you? We shall charge you at Lambeth as a receiver of stolen goods: you will be remanded for a week in custody, and by that time we shall have your Prince in safe keeping in St Petersburg."

"Will you, really?" asked Phadrig, lifting his eyelids for the first time during the interview. "I should have thought that a man of your European experience would have called the Russian capital by its proper name. Surely you know that only newspaper people make that mistake. It is the city of Peter the Great, not Saint Peter the apostle. The fortress of Petro-paulovsky is not named after saints—only after Tsars."

There was a sneer in his voice as he made this trivial correction which roused both Hendry and Von Hamner to anger. The German pulled his revolver out of his hip pocket, and Hendry produced a beautiful pair of polished handcuffs from his left trouser pocket.

"Ah, I see that you have come prepared, gentlemen!" said Phadrig, with a laughing sneer in his low-voiced whisper. "Those are what you call the bracelets in England, are they not? Well, since you are determined to take the law into your hands—here are mine. Put them on M. Hendry, and then your friend may not think it necessary to try and shoot me."

He held his hands out. The way in which he said "try and shoot me" did not sound well in their ears, but Nicol Hendry thought that the work had to be put through now or not at all. He took a couple of steps towards Phadrig, and a couple of sharp snaps told Von Hamner that their prisoner was safe. But the prisoner did not seem to think so. He raised his hands and looked at the handcuffs. He seemed to examine them as though they were curiosities.

"Are these really what you take criminals to prison with? They don't seem very strong. I could break them as though they were thread."

"That will do, Mr Amena. You've got them on now, and we don't want any more of your conjuring tricks. Come along, and take it quietly like a sensible man."

Hendry was fast losing patience, and Von Hamner was doing all he could to keep his finger off the trigger of the revolver.

"Ah yes, conjuring tricks you call them, you ignorants! Now look. You have put the handcuffs on to my wrists. Is this a conjuring trick? See!"

He held his arms out towards them, his two hands chained together.

"Mr Hendry, be good enough to take my right hand, and you, Herr von Hamner, my left. So; now shake my hands. You see, there are the handcuffs on the floor."

It was only a shake of the hands, but the clink of the steel followed as the bracelets dropped from his wrists. He stooped down, and inside ten seconds they were clipped round Von Hamner's. In the same instant he had twitched the revolver out of his hand and pointed it at Hendry's face.

"Now, gentlemen, you were talking about taking the law into your own hands. I, you see, have taken it into mine. What do you propose to do? I am quite at your service. Your idea of arresting me on a charge of receiving stolen goods is, if you will allow me to say so, absurd. You could no more make me guilty of that than you could hang me for the deaths of those foolish spies of yours. Now, what is it to be? Pardon me, Herr von Hamner: the bracelets inconvenience you. Allow me." He took the handcuffs between his finger and thumb, shook the chain, and they dropped into his hand. "You will feel more comfortable now."

"Yes, and I'll make you less comfortable in Hell, where you should have been long ago," shouted Von Hamner, jumping at him the moment his hands were free, and snatching the revolver out of his hand. The pistol went up before Hendry could get hold of his arm, and he fired. Phadrig put his hand up, and when the smoke had drifted away, he held it out to Von Hamner, and said:

"I think that is your bullet, mein Herr."

The bullet was lying in the palm of his hand, a little out of shape through passing the rifling, but still the same bullet.

The German's face turned a reddish-grey, and Nicol Hendry, with all his courage, was not feeling particularly well. As a matter of fact, he was, for the first time in his life, absolutely frightened. A man who could deal with handcuffs as though they were made of cotton, and catch a bullet in his hands, was not the sort of criminal he had been trained to hunt. As for Von Hamner, he was in a state of utter collapse. He dropped upon a chair, a pitiable spectacle of craven fear, looking about half his real size so physically shrunken did he seem.

"Let the devil go, Hendry," he mumbled. "He is more than man. What is the use? If you cannot shoot him, you cannot hang him, and if handcuffs won't hold him, prison doors won't. Let us go and leave the devil to himself. I've had enough of it."

"But perhaps the devil has not," said Phadrig, with a politeness which was infuriating in its mildness. "You gentlemen will understand that I do not wish to have this espionage going on any longer. If you cannot promise that it shall stop at once I shall, for my own protection, have to suggest to you that you shall remove yourselves, as the others have done."

"No, no, not that, man, not that!" shouted Von Hamner, springing from his seat and making for the door. "I have done with the whole business, curse it! Let me go, let me go! Hendry, do as you like, but do it alone. I have finished."

Before Hendry could reply, or before Von Hamner could reach it, the door was flung open, and Franklin Marmion strode into the room. Von Hamner crawled back to his chair. He did not like the look of a dead man who had come to life again. Nicol Hendry held out his hand, and said:

"And is it really you, Professor? Mr Amena here has just had news that you were dead—'fallen overboard in the Baltic from Prince Oscarovitch's yacht. Body not recovered,' is what the telegram says."

"The body is here right enough, M. Hendry. I did not fall overboard. I was bound hand and foot, had a mass of iron tied to my feet, and was thrown out of a port-hole by the Prince and his captain. Of course, I got rid of the rope and the iron even more easily than this man got rid of your handcuffs a short time ago, and after keeping myself afloat for half an hour or so, I was picked up by a fishing-boat which took me to Stralsund. I got a change of clothes there, and came home viâ Hamburg and Ostend. My daughter has gone on in the yacht to Oscarburg, where the Prince expects to make her his wife, and where she will make a very considerable fool of him. That is all, and now I suppose I had better deal with this man."

"Mercy, mercy, Thou Who Knowest! Pity, pity!"

Phadrig raised his hands above his head, turned round thrice slowly, and sank in a heap on the floor.

"Thou who wast once High Priest in the House of Ptah: thou who hast held the Doctrine: thou darest to ask for mercy, knowing well that there is no forgiveness of sins: thou hast taken innocent lives, believing thyself above human law. A wasted life is behind thee: see that thou doest better for thy soul's sake in the next. Die now! The High Gods have spoken, and the penalty of sin is death—and the life beyond. Die!"

And Phadrig died. His eyes glazed and his flesh withered; his lips and his gums dried up and shrivelled away from his jaws. His clothes fell away from his body in rotting shreds, and before Nicol Hendry and Von Hamner had quite grasped the full meaning of the horror that was happening before their eyes, all that was left of him was a little heap of yellow bones with a few fragments of cloth clinging to them.

"Gentlemen," said Franklin Marmion, "there are some things which cannot be told. I think you will agree with me that this is one of them. Mr Amena has left the world for the present. Those bones will be dust in a few minutes. It will only be another mysterious disappearance, and I don't think that any one except the Pentanas and Prince Oscarovitch will trouble much about him. The Pentanas are now deprived of all power for harm, and the Prince will probably be a harmless lunatic when he comes back into the world. I should sweep that dust up and put it into the fireplace, if I were you. In that desk you will find documents giving the whole history of the Affaire Zastrow. They will be useful to you. You will have to excuse me now. Europe is on the brink of war, and I must go and remove the cause. I rely upon your discretion as to the events of this afternoon. Au revoir. I shall have the pleasure of seeing you again shortly."

The door closed, and they were left to their somewhat gruesome task.

CHAPTER XXVI
CAPTAIN MERRILL'S COMMISSION

Franklin Marmion found a hansom in the Borough Road and drove to Waterloo. He had just time to wire to Merrill to meet him at the "Keppel's Head" for dinner and catch the new 4.55 express for Portsmouth. Merrill was waiting for him in the smoking-room. As they shook hands, he said in the quiet tone which is characteristic of his profession:

"Your wire was rather sudden news, Professor. I thought you were somewhere in the Baltic. Your coming back like this seemed to mean something, and so I took the liberty of having a private room for our dinner."

"Perfectly right, my dear Merrill," he replied. "Let us go upstairs at once. I have a good deal to say to you, and what I am going to say will have to be done quickly."

"We have our sailing orders for the Baltic, and the Special Squadron leaves Spithead at midnight. Come upstairs, Professor, and we can talk."

Dinner was served a few minutes after they got into the room that Merrill had reserved on the first floor. The waiter was dismissed and the door locked, and then Franklin Marmion told Mark Merrill the most wonderful story he had ever heard. If it had come from any one else he would have put it down as a lie, but he remembered what had happened in the lecture theatre of the Royal Society, and so he held his peace. It was quite impossible for him to disbelieve anything the father of his Best Beloved told him. When the Professor had finished the story of Nitocris and the Prince, he leaned his elbows on the table, and said:

"Now, my dear Merrill, I am going to put it into your power to save Europe from the horrors of a universal war: but to that you must be prepared to take risks which may result in your being dismissed the Service. On the other hand, if you succeed, as you are almost certain to do if you act strictly on the instructions that I am going to give you, you will be a Captain in a month, and a Vice-Admiral in a year."

"But I'm a Captain now, Professor. I was keeping that little bit of news for you. I hoisted my pennant this morning on His Majesty's ship *Nitocris*: new second-class cruiser, eight thousand tons, and twenty-four knots: as pretty a ship as Elswick ever turned out. And the name: it came to me like a revelation."

"Possibly it was, in a sense that you may not quite understand now, but you will understand it when you and Niti are married. She will be better able to explain it then than I could now."

"And what are the orders—I mean, of course, the private ones? Ours are: sail at midnight, make Kronstadt in forty-eight hours: command the approaches to Riga and St Petersburg, and wait for the developments of this manifesto which seems to be setting what is left of Russia on fire. Germany is in with us for the time being: France and Italy and our Mediterranean squadron will see to things in the Near East, and altogether there seem to be the prospects of a very handsome sort of row."

"Which you, my dear Merrill, will be the means of preventing," said Franklin Marmion, taking a piece of folded tracing paper out of the inside pocket of his coat. "I yield to circumstance. The name of your new ship convinces me that I was wrong in certain other circumstances. You will give me a passage to Viborg on the *Nitocris*. You will take French leave of the fleet as soon as you sight Kronstadt, get into Viborg Bay at your best speed, land your men, take the Castle, which is quite undefended, bring away Prince Zastrow and Oscarovitch, and, of course, Niti; put your two princes on board the flagship, bring them back to England, and dictate terms from London. It seems a good deal to do, but I will make it possible, if you are prepared to do as I advise you. There is the chart showing the approaches to Oscarburg."

"I'll do it, sir," said Merrill, taking the tracing from his hand. "I'll break every regulation of the Service into little pieces to get that done. Now, I ought to be getting on board. Are you ready?"

"Quite," said Franklin Marmion, rising from his chair. "I see now where the man of action comes in. I did not see that before, I must confess."

CHAPTER XXVII
THE BRIDAL OF OSCAROVITCH

The Special Service Squadron steamed out of Spithead as the clock of Portsmouth Town Hall chimed twelve that night. Thirty-six hours later a marriage ceremony took place in the chapel of the Castle of Oscarburg. It was performed according to the rites of the Orthodox Church, and the witnesses were Prince Zastrow and his medical attendant, Doctor Hugo. The retainers of the Castle, headed by the major-domo and the housekeeper, formed the congregation. Jenny was up in her mistress' room packing as though for an immediate departure. She was very frightened at the happenings of the past three or four days, but she contented herself with the thought that her mistress was going to be a princess, and that, therefore, her own lot in life would be brightened with reflected glory.

When the ceremony was over, the wedding feast was held in the great dining-hall of the Castle after the ancient Finnish style. When the loving-cup had been drunk, Nitocris took leave of her lord and went to her room. The bridal chamber was blazing with light, and the great silken-hung bed was a couch fit for a queen. She turned the draperies down, laid herself dressed on the thick, downy bed, and then got up and went back to her own.

"I shall sleep here to-night, Jenny, and I shall not undress. You mustn't do, either. Lock the door, and put the sofa across it. You will find that something is going to happen to-night. Is everything ready for us to go away?"

"Yes, Your Highness," replied Jenny, wondering what was going to happen next.

"You must not call me Highness, Jenny," said her mistress, with a laugh. "I did not marry the Prince to-day. It was some one else he knew a long time ago. I have put her to bed in that splendid bridal chamber of his. She is waiting for him now."

"But I don't understand, Miss—I——"

"There is no need for you to understand, Jenny. Just be a good girl, and do as you're told. When we get back to England I will explain matters as far as I can."

Miss Jenny wisely decided to keep her thoughts to herself, and went on with her packing. Nitocris changed her bridal dress for her yachting costume, and lay down on the couch to await the progress of events.

Oscarovitch left the company in the dining-hall to their revel in about an hour's time, and went up to his fate in the bridal chamber. He knocked and opened the door softly: locked it, and went toward the bed. He leaned over it for a moment, and then a hoarse shriek of mingled rage and terror rang through the room. He flung the clothes off the bed. Where was the lovely bride he had wedded only a few hours before? What was this horrible thing lying where *she* should have been? Not Nitocris—and yet, it *was* Nitocris. Like a flash of lightning rending the darkness of the midnight heavens, the gap of oblivion between his lives was rent, and the light flamed into his soul. Phadrig had lied to him. The daughter of Rameses had not died that night in the banqueting chamber of the Palace of Pepi. She had lived and reigned virgin queen of the Sacred Land. Her body had been submitted to the hands of the paraschites and buried in the City of the Dead over against Memphis, on the eastward side of the river. And here was her mummy lying in his bridal bed, mocking him with its hideous, stony rigidity.

For a few terrible moments he stood staring at it, his clenched fists raised above his head. Then with another scream he cast himself upon it.

When they broke the door open, they found the man who in a few days would have been Emperor of the Russias and the East lying across the bed mowing and gibbering like a mad monkey, and scraping up handfuls of brown dust from the stained sheets.

Twenty-four hours later the Admiral in command of the British Special Squadron off Kronstadt saw the private signal flashed from the north-east. He was a very angry Admiral, for he had lost a brand-new cruiser and one of the smartest captains in the Service. But the signal spelt "*Nitocris*. All well. Coming alongside."

"All well, and be damned to you, Captain Merrill!" muttered the Admiral under his breath, when the signal was read to him. "This is a nice way to begin a new command. I've half a mind to put him under arrest: but he's a good man. I'd better hear what he has to say for himself first. I wonder what the deuce he's been doing with that cruiser since he took her away without leave? Well, here she is, I suppose."

But it was not H.M.S. *Nitocris* that came out of the night glittering with electric lights and flying through the water at a speed that the fastest destroyer in the squadron could not have equalled. A whistle tooted softly, a white shape swung up out of the darkness and slowed down alongside the flagship. A boat dropped into the water, and three minutes later Captain Mark Merrill ran up the gangway ladder, saluted the quarter-deck, and handed his sword to the Admiral.

"I have done wrong, sir, but I hope that I have also, in another sense, done right. I have brought both princes with me."

"Both princes—Good Lord, sir, what do you mean?"

"May I come below with you, sir, and explain? It has been rather delicate work, but we've got it through all right, I think."

"Then keep your sword for the present, and come and tell me what you have to say."

Captain Merrill followed the Admiral to his room, and told the story of the taking of the Oscarburg—a very easy matter with a hundred bluejackets at his back—the capture of Oscarovitch, who was now in a straight waistcoat on board his own yacht, the rescue of Prince Zastrow and Nitocris, and — —

"The other Nitocris is following, sir," he concluded. "I thought I had better take the yacht. She can make a good thirty-five knots, and that's useful when you're in a hurry. And now, sir, I am at your disposal."

"Rubbish!" said the Admiral, holding out his hand. "Captain Merrill, I don't quite know how you've done it, but you've saved Europe, and perhaps the world, from war. If you hadn't brought those two princes of yours to-night, we should have been fighting Germany for the possession of Kronstadt before mid-day to-morrow. Those were the orders. Now, of course, they can do nothing, as you have brought Prince Zastrow back from the dead. He's their choice, and you had better get him and the other away to London as soon as I have seen them, and you can take my report with you on that thirty-five knotter after breakfast to-morrow morning. Now, it's getting late. I'll say good-night."

EPILOGUE

The double wedding which took place at St George's, Hanover Square, the following June was one of the most brilliant functions of the year. Their Majesties of Russia and Great Britain graced the ceremony with their presence, and, as a special act of grace to the man who, with Franklin Marmion's help, had saved the world from what might have been one of the bloodiest wars in history, H.M.S. *Nitocris* was put into commission for a cruise, the object of which was anything rather than warlike. Two of the happiest couples on land or sea made the round of the world in her. Before they returned Princess Hermia had taken the last of Phadrig's drug and lain down to sleep never to wake again, and in the fullness of her happiness Nitocris pardoned Oscar Oscarovitch, and allowed him to die.